W9-AVR-720

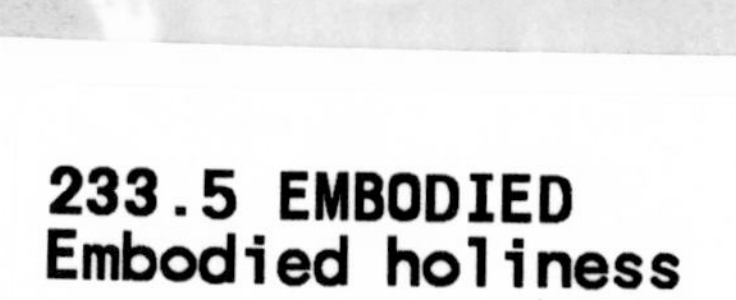

Toward
A Corporate Theology
of Spiritual Growth

Embodied Holiness

Edited by Samuel M. Powell
& Michael E. Lodahl

InterVarsity Press
Downers Grove, Illinois

InterVarsity Press
P.O. Box 1400, Downers Grove, IL 60515
World Wide Web: www.ivpress.com
E-mail: mail@ivpress.com

InterVarsity Press® is the book-publishing division of InterVarsity Christian Fellowship/USA®, a student movement active on campus at hundreds of universities, colleges and schools of nursing in the United States of America, and a member movement of the International Fellowship of Evangelical Students. For information about local and regional activities, write Public Relations Dept., InterVarsity Christian Fellowship/USA, 6400 Schroeder Rd., P.O. Box 7895, Madison, WI 53707-7895.

Cover photograph: William Koechling

ISBN 0-8308-1583-X

Printed in the United States of America ∞

Library of Congress Cataloging-in-Publication Data has been requested.

23 22 21 20 19 18 17 16 15 14 13 12 11 10 9 8 7 6 5 4 3 2 1
18 17 16 15 14 13 12 11 10 09 08 07 06 05 04 03 02 01 00 99

Contents

Introduction

Samuel M. Powell

The phrase "embodied holiness" seems to be an unnecessary and uneconomical redundancy, at least if it is supposed to describe the people of God. After all, what would a disembodied holiness be? Of course, God's holiness would be of the disembodied sort, as would that of angels; however, our subject is human holiness. And a purely spiritual holiness that left behind the body would not be holiness at all but something else.

So why should we need a discussion of *embodied* holiness? Why not simply another book on *holiness,* without corporeal qualification? Indeed, has any Christian ever actually advocated a disembodied holiness? With the possible exception of the Gnostic Christians, who are commonly thought to have rigidly separated soul from body, it has always been acknowledged that holiness is a bodily as well as a spiritual matter. Monks rigorously guarded speech and physical labor, mystics suffered the wounds of the cross, and the faithful consumed the body and blood of Christ. Why then is there today a call for embodied holiness, a call like a prophetic oracle summoning God's people back to a lost heritage?

That a concern with things embodied is a part of the heritage of God's people is evident from the various meanings of *body* in Christian thought. These meanings determine what "embodied holiness" can mean. "Body" in Christian thought commonly signifies the physical body of the individual believer; the physical (but then resurrected) body of the historical Jesus; the church, of which Christ is the head;

and (at least in some traditions) that which is consumed (in ways variously understood) in the Lord's Supper. These meanings are all connected with each other. The body that is consumed in the Lord's Supper is the resurrected and glorified body of Jesus Christ, and it is this same body that is the church. Further, membership in this body of Christ (1 Cor 12:27) and participation in the body of Christ (1 Cor 10:16) imply something about the holiness of the individual Christian's body (1 Cor 6:19-20). So, the physical bodies of believers are clearly subjects of holiness, as the many exhortations of Paul's letters remind us. The physical body of Jesus Christ is of capital importance in holiness, both because of its central role in the doctrine of the atonement and because of the exemplary value of Jesus' life for Christians. The church as the body of Christ looms large in any discussion of holiness, not only because it is the church as such that Christ wishes to make holy, without spot or wrinkle, but also because the church is the living context in which the Christian becomes holy. Finally, the sacramental body of Christ is important for holiness because thereby Christ is incorporated into us and we into him. In short, the centrality of the body to the concept of holiness is plainly a teaching of the New Testament.

Yet all this has been known for centuries, even if it has not had as great an impact as we may wish. We still do not know why "*embodied* holiness" should have the urgency of a prophetic oracle. To see why it does have this urgency today and why it did not have such urgency in the past few decades and centuries, it will be helpful to take note of the leading essay in this book, Stanley Hauerwas's "The Sanctified Body: Why Perfection Does Not Require a 'Self.'" This essay was delivered as an address at a conference sponsored by the Wesley Center for Twenty-First Century Studies at Point Loma Nazarene University in January 1997. The theme of the conference was "What Happens to 'Person' in a Postmodern Era?" This theme alludes to the fact that the concept of the self has become problematic for us; the title of Hauerwas's address suggests that with respect to holiness the body is a more relevant consideration than the soul or self. If this theme and this title are indicative, then a momentous intellectual change is upon us, one that touches on our understanding of human being. The issue in "embodied holiness" consequently is not whether or not there can be a purely spiritual holiness but instead whether there is anything beside the body that can be sanctified. Do terms such as *person* and *self* denote anything in addition to our bodily existence? We thus find ourselves in a Christian intellectual milieu in which positivist and physicalist ways of thinking are recommended for Christian thought in spite of their odious and disreputable association with the Logical Positivist movement of the 1930s and 1940s. The church has a way of needing a generation or two in order to become accustomed to philosophical ideas.

It seems, then, that the prophetic urgency lying behind "embodied holiness" is generated by twentieth century doubts about the nature of the self and the self's relation to the body. Evidence that fascination with the body is not a peculiarity of holiness thinkers but is rather a widespread cultural phenomenon is easily found. Logical Positivism's physicalist theories have already been mentioned. We could also look to the philosophy of Ludwig Wittgenstein, with its rather behaviorist theory of language. However, there is a more proximate example, namely those French philosophers often identified with the poststructuralist way of thinking. They make a prominent use of the concept of the body. For example, consider Michel Foucault, who writes:

> In our societies, the systems of punishment are to be situated in a certain "political economy" of the body. . . . It is always the body that is at issue. . . . The body is also directly involved in a political field; power relations have an immediate hold upon it. . . . The body becomes a useful force only if it is both a productive body and a subjected body.[1]

Compare this with Hauerwas's claim that "Christianity is to have one's body shaped, one's habits determined, in [such] a manner that the worship of God is unavoidable" (p. 22). This statement implies that self-determination, always a concern to the upholders of the

---

[1]Michel Foucault, *Discipline and Punish: The Birth of the Prison,* trans. Alan Sheridan (New York: Vintage, 1979), pp. 25-26.

soul-concept, is of little importance. The self that determines itself is to be replaced in our thinking by the body that is to become subject.

Or, attend to what Foucault says about the object of punishment in the modern world:

> What one is trying to restore in this technique of correction is . . . the obedient subject, the individual subjected to habits, rules, orders. . . . The training of behaviour by a full time-table, the acquisition of habits, the constraints of the body imply a very special relation between the individual who is punished and the individual who punishes him.[2]

In a similar vein, Hauerwas asserts that "I purposefully use the language of 'force' because I think that we do not develop new habits without being forced to do so. If we are to learn to think differently, we must have our bodies repositioned so that we have no other choice but to be what we were created to be" (pp. 35-36). We note the at least superficial points of resemblance between Foucault's and Hauerwas's language. Of course, Hauerwas is not simply adopting Foucault's analysis in its entirety; Foucault after all was not a Christian. Nevertheless, what is remarkable is the extent to which both Foucault and Hauerwas share a common vocabulary, a vocabulary that gives primacy to the body when the nature of human being is considered.

Hauerwas imparts this primacy to the body in an attempt to escape from the long shadow of Rene Descartes's dualistic metaphysics of soul and body, a dualism that is often credited with being both the beginning of and also the curse of the modern world. Along with Hauerwas, Craig Keen sets himself against this Cartesian dualism, lamenting its "preoccupation with the highly abstract notion of the self," a self that is an "*I* who stand[s] as the single indubitable truth, the absolutely solid rock foundation of everything that might be judged to be true" (p. 46). In place of this solitary self of modern philosophy Keen wishes to substitute a Trinitarian and communitarian understanding of the self. The burden of both Hauerwas's and Keen's essays is the rethinking of the concept of the self in such a way that

---

[2]Ibid., pp. 128-29.

the Promethean modern self is bound once again to the rock, where it may be consumed, and in such a way that a more social, contextual and above all corporeal notion of the self emerges. Hauerwas and Keen thus number themselves among the class of postmodern thinkers. Such thinkers distinguish themselves by their steadfast resistance of the pernicious influence of Descartes's understanding of the self. Indeed, they regard Descartes as both the progenitor and the apotheosis of the modern era in philosophy. Consequently, being post- (really anti-) Descartes signals being postmodern. But in fact it may be simpler to consider this class of thinkers more post-Cartesian than postmodern, for the body as a central theme of philosophy is hardly a recent invention. The materialistic philosophies of eighteenth-century France are well-known, as is the empiricism of David Hume, whose doubts about the self rival those of any postmodern philosopher. But even in the modern tradition of metaphysics we find a sustained and appreciative consideration of the body. For example, Gilles Deleuze claims the following of Spinoza:

> Spinoza offers philosophers a new model: the body. He proposes to establish the body as a model: "We do not even know what the body can do. . . ." What does Spinoza mean when he invites us to take the body as a model? It is a matter of showing that the body surpasses the knowledge that we have of it.[3]

In short, Spinoza accomplished the philosophical task of grasping the nature of mind and body without recourse to the illicit concept of consciousness. Along similar lines, Rodney Clapp urges us to consider Michael Polanyi's notion of tacit knowledge and to apply this notion to holiness. Such tacit knowledge exhibited, for example, in the sacraments, is more practical than theoretical, more a matter of the body than of consciousness. Here something is known without the participant being expressly conscious of that knowledge, just as a child knows how to ride a bicycle without knowing the physical principles involved (pp. 64-65). Why make use of the idea of a soul when the

---

[3]Gilles Deleuze, *Spinoza: Practical Philosophy*, trans. Robert Hurley (San Francisco: City Light Books, 1988), pp. 17-18.

powers of the body in embodied knowledge and embodied holiness lie untested and unexplored?

Hauerwas's strictures against the modern view of the self and his lifting up of the body, Keen's critique of Descartes and appeal to the Trinity as a model of human being, and Clapp's opposition of tacit knowledge to consciousness, then, are episodes in a general philosophical revolt against the branch of modernism that derives from Descartes in favor of another branch of modernism that has affinities with Spinoza and a great many other moderns. The urgency of embodied holiness is a function of the split personality of modern thought, the Cartesian and the Spinozan in an unresolved conflict. Ironic that, while the body of Christ is one, the self of modernity is split, its soul rent.

That this preference for the body and suspicion of the concept of the soul is a widespread phenomenon in theology is evident from such books as *Whatever Happened to the Soul?*[4] This is a recently published anthology whose aim is to argue that Christian theology does not require the concept of a soul (in the dualist sense), that modern neurobiology does not permit such a concept and that Christians should adopt what the authors call "nonreductive physicalism." These authors are thus arguing for the same point that Hauerwas is, namely that the concept of the soul is a menace to good theology and that attention is better paid to understanding the ways in which human life is embodied. Attention could also be drawn to the various exponents of process theology, whose quarrel with Cartesian dualism is well-known and to the various sorts of liberation theologies, with their strong commitment to contextualized theology and to the social-political effects of theology.

We can see, then, that both Christian theologians and non-Christian philosophers have lately discovered the demerits of dualistic views of human being and have championed the cause of the body. These acts of discovering and championing are the contemporary cul-

[4]Warren S. Brown, Nancey Murphy and H. Newton Malony, eds., *Whatever Happened to the Soul? Scientific and Theological Portraits of Human Nature* (Minneapolis: Fortress, 1998).

tural current that has generated the sense of prophetic urgency about the importance of the body. This cultural milieu has enabled and encouraged theologians to reach back into the Christian tradition and once again to emphasize the corporeal aspects of the Christian life.

But there is a second impetus for the theme of embodied holiness, one that bears directly upon the essays in this book. That source is the Wesleyan theological heritage. All the contributors to this book share that heritage. Accordingly, it may be helpful to indicate why a book has been assembled using a Wesleyan cast of characters and how Wesleyan theology relates to the theme of embodied holiness.

First, a disclaimer: a book of essays by *Wesleyans* on the subject of holiness does not presuppose or imply that only Wesleyans have an understanding of holiness or even that Wesleyans have the best understanding of holiness. On the contrary, it signals that holiness has become a problematic topic in Wesleyan circles. As is well-known, John Wesley's theology made a prominent place for the doctrine of Christian perfection, better known in the American Wesleyan tradition as holiness. What may not be as well-known is that Wesleyans have been as contentious about how to define holiness as they have been reticent to actually pursue holiness. As a result, the Wesleyan tradition has witnessed repeated, unedifying arguments of the *who-has-the-correct-view-of-holiness?* variety. Although this volume does not claim to have offered the final word—not even the final Wesleyan word—on the subject of holiness, it does represent a step forward by trying to come to terms with John Wesley's doctrine of perfection. This step is not in the direction of a greater consensus, as though all the contributors to this book were in complete agreement. However, it is an attempt to move the church's understanding of holiness beyond some of the more labored interpretations of the past and toward an engagement with the pressing issues of the contemporary situation.

As to the connection between Wesleyan theology and embodied holiness: First, there is John Wesley's own insistence on the practical character of holiness. By this is meant that holiness is a matter of practice. It is not merely a status that one attains upon justification or some other moment; it is instead something that is lived and that shapes our lives. For John Wesley the practicality of holiness can be seen in the disciplines that lead up to holiness and also in the ethical consequences that result from holiness. Christian perfection is not, for Wesley, something given to us in the way that a letter is received in the mail; instead it is formed within us as we put to death the deeds of the body by the practice of Christian disciplines. We are not to wait for it in mystical stillness but in an active striving. Further, the striving for and attaining of Christian perfection issues forth in a host of good works characterized by love of neighbor. This practical aspect of holiness, then, is one sense in which holiness is to be embodied. It is not only a matter of prayer, meditation and subjective awareness, but is also the doing of God's will in all forms. To illustrate this point, Joyce Quiring Erickson's essay examines the lives of prominent Methodist women in the eighteenth century by studying their diaries and records. In these the routines and practices of spiritual disciplines are recorded as they trod the path of holiness. Erickson suggests that such women and especially the disciplines that they willingly gave themselves to may even today function as a model for us.

Second, "embodied" can also be taken to imply the ecclesial character of the Christian life, the fact that there can be no Christian life apart from the life of the church. Here again John Wesley's own theology still has something valuable to say to us. As Wesley himself put it, there is no holiness except social holiness, by which he meant a holiness that occurs in and through the church. Nor was he squeamish about specifying exactly what it is about the church that is instrumental in leading to Christian perfection. In his view worship, attendance at the sacraments and other churchly functions are all essential to the Christian life. Embodied holiness, then, means a holiness that can be attained only in and through the ecclesial body of Christ. In connection with this theme the essays by Theodore Runyon and Michael G. Cartwright seek to expound the social or communal character of holiness. Runyon's essay focuses on Wesley's understanding of the image of God. Rather than a static possession, the image is, Runyon maintains, the basis for seeing holiness as essentially communal. Cartwright's essay is an analysis of the American religious classic *In His Steps.* His thesis is that the American Protestant pursuit of holiness is marred by the failure to engage in this pursuit within the context of the church's disciplines and practices. Only in the church he argues, do we find the means for effectively attaining the life of holiness; without these means we pursue only empty phantoms of our own making.

This second aspect of Wesleyan theology (that embodied holiness is social holiness) is the obverse of concern about the dangers of the Cartesian self. Further, interest in the social nature of Christianity is no possession of Wesleyan thought alone but is in fact a central concern of a particular branch of modern theology, namely liberal theology. Although commonly represented otherwise, many liberal theologians were intent on affirming the social nature of the Christian life over against modern, liberal, political individualism. Hence, the struggle against an excessively individualistic view of the Christian life, which is attested in all the essays in this book, must be regarded not as a reaction against modern thought but as one stream of modern thought wrestling with another stream. Accordingly, while an emphasis on individualism is certainly a prominent aspect of some modern thought, it would be a mistake to regard it as the only modern view of human being. For example, already in the nineteenth century liberal theologians had conceived of Christianity in organic terms similar to those described by the essayists in this book. Albrecht Ritschl, complaining of scholars who sought to write an objective history of Jesus without presuppositions, argued that "If we can rightly know God only if we know Him through Christ, then we can know Him only if we belong to the community of believers."[5] In other words, the only way of grasping the true significance of Jesus is by being a member of the church, for only there is the purpose of Jesus fulfilled.[6] Earlier in the century Friedrich Schleiermacher had argued that the

[5]Albrecht Ritschl, *The Christian Doctrine of Justification and Reconciliation: The Positive Development of the Doctrine,* trans. H. R. Mackintosh and A. B. Macaulay (New York: Charles Scribner's Sons, 1900), p. 7.

[6]Ibid., p. 2.

church is a sort of new humanity in which alone is the influence of Christ effective. One is, for Schleiermacher, a Christian only as one is incorporated into this new humanity. For example, he asserted that "We are all conscious of all approximations to the state of blessedness which occur in the Christian life as being grounded in a new divinely-effected corporate life."[7] Such a view, he maintains, excludes

> the idea that one can share in the redemption and be made blessed through Christ outside the corporate life which He instituted, as it were, alone. This separatism disregards the fact that what owes its origin to divine agency can nevertheless be received only as it appears in history, and also can continue to function only as a historical entity. . . . It destroys the essence of Christianity by postulating an activity of Christ which is not mediated in time and space.[8]

These influential representatives of liberal theology show that modern thought includes a trajectory whereby Christianity is understood in thoroughly historical and social terms. Of course, the claim here is not that Hauerwas and others asserting the social nature of Christianity are simply reverting to an outdated form of liberal theology. However, it is fair to state that the postmodern emphasis on the social character of holiness and the ecclesial character of the Christian life are developments of this thoroughly modern trajectory. Consequently, just as the reaction against Cartesian dualism that one finds in this collection of essays represents one stream of modernist thought (exemplified in Spinoza) in its struggle against another (initiated by Descartes), so the emphasis on Christianity's social nature pits one branch of modern thought (the liberal theological) over against another branch (liberal individualism).

However, in spite of the truth to be recovered from this renewal of communitarian thought, the danger looms of connecting holiness to the church in such a way that the Christian life becomes ingrown and the church finds its center in itself. The church could become an

---

[7]Friedrich Schleiermacher, *The Christian Faith*, trans. H. R. Mackintosh and J. S. Stewart (Philadelphia: Fortress, 1976), p. 358.
[8]Ibid., p. 360.

enclosed in-group that rigorously separates itself from the world in such a way that a dualism arises—not a dualism of soul and body but a dualism of insiders and outsiders. Accordingly, the essays by Keen and Michael E. Lodahl expound holiness as a movement from the inside of the church to the outside. Keen judges Hauerwas wanting in just this respect. He finds troubling Hauerwas's use of the language of naturalism and necessity, with it implication of a closed system, and Hauerwas's implied emphasis on community identity, with its clear demarcation between insiders and outsiders. To the same end Lodahl employs the parable of the good Samaritan to argue for an inclusive understanding of the church. The story, Lodahl argues, is about holy love and its role in transgressing such boundaries as Jew and Samaritan and the pure and the impure.

Another and related threat is that replacing the individualistic view of the Christian life with the social view will simply substitute one sort of humanism for another. Instead of Christianity being interpreted in terms of individual self-consciousness, it would be interpreted in terms of communal consciousness. As a response to this danger I engage the thought of Friedrich Schleiermacher and Karl Barth to devise an understanding of holiness that will acknowledge the genuinely human and social dimensions of the Christian life while insisting that such a life has its center in God. Arguing against a premature identification of holiness with ethics, I urge readers to consider holiness as lying in our encounter with God, an encounter that then results in an ethical love of neighbor.

The notion of embodied holiness is both timely and timeless. On the one hand, it arises out of sensitivity to our contemporary situation in which the fundamental conceptions of human being and community are being reassessed and revised. On the other hand, it grows out of the command of God to love. Holiness is always contextual but never relative, always embedded in our situation but never merely situational. In short, holiness is a practice in which our doctrines are corporealized, a doctrine as closely related to practice as the soul is to the body. The essays in this book are all invitations to consider holiness both as a practical doctrine and as a doctrinal practice.

1

The Sanctified Body

Why Perfection Does Not Require a "Self"

STANLEY HAUERWAS

I am an evangelical Catholic—which is but a way to say that I am a Methodist. Methodism is a movement that by accident became a church, or at least claims to be a church, in America. John Wesley charged his followers to cover the world. We have done that, but as a friend observes, like Sherwin-Williams paint we are only a fourth of an inch deep: a strange result for a holiness movement meant to renew an established church. Yet at least on some readings Wesley represented a peculiar combination of Catholic and perfectionist theological convictions.[1] As one wag once put it, "If Wesley had not

[1]I do not mean to suggest that sanctification and holiness are not equally central to Catholic practice and theology. The crucial issue is where such holiness is to be found. Geoffrey Wainwright in *Methodists in Dialogue* ([Nashville: Kingswood Books, 1995], p. 20) helpfully notes that Catholic teaching serves to give the believer an assurance concerning the church's proclamation of the gospel, while the Methodist tends to point to the assurance of the Spirit manifest in the individual believer. Wainwright quotes from the 1986 Nairobi Report of the Methodist-Roman Catholic Joint Commission that rightly suggests that Methodists "might ask whether the Church, like individ-

been Wesley he would have been Ignatius Loyola." Or, more seriously, there is Albert Outler's frequent claim that the true character of Methodism is to be an "evangelical order" with "a catholic Church."[2]

I have always found this peculiar combination, even though it may not represent the "historical Wesley," theologically attractive and promising. Yet one must also acknowledge that the position is inherently unstable, particularly in modernity. It will be the burden of my argument that recent developments, which some call postmodernism, offer some extremely helpful ways for a display of holiness without a loss of the catholic character of the church. The loss of the "self" and the increasing appreciation of the significance of the body—and in particular the body's permeability—can help us rediscover holiness not as an individual achievement but as the work of the Holy Spirit building up the body of Christ.

Yet I first need to suggest why an evangelical Catholicism has been so hard to sustain theologically as well as in the actual practice of the church in modernity.[3] To do so I am going to make generalizations about Roman Catholicism as well as Protestant practice that would require much qualification if they were to reflect the complexity of each. I think it worthwhile to risk being accused of overgeneralization

---

uals, might by the working of the Holy Spirit receive as a gift from God in its living, preaching and mission, an assurance concerning its grasp of the fundamental doctrines of the faith such as to exclude all doubt, and whether the teaching ministry of the Church has a special and divinely guided part to play in this" (p. 21). That question asks Methodists to develop an adequate ecclesiology that, given our beginnings, has always been absent.

[2]Wainwright uses Outler's statement in a sympathetic way throughout his book. As Wainwright observes, "If one were to reflect more fully than John Wesley did on the theological implications for ecclesiology of [his] view of the appropriation of salvation, it would put Wesley rather on the 'Catholic' side in the debates concerning the instrumentality of the Church, in which some contemporary ecumenists have located the basic difference between Roman Catholics and Protestants" (*Methodists in Dialogue,* p. 104). That seems exactly right.

[3]Pietism and holiness, while often connected, customarily are not only distinguishable but can be antithetical. That is particularly the case when the emphasis of pietism on the individual's experience undercuts any account of the importance of discipline as a form of discipleship. One of the ironies of American Methodism is how the revivalistic enthusiasm of the nineteenth century served to render impotent Wesley's understanding of these matters.

in the interest of making clear the challenge I see before us. By "us" I mean those who think any intelligible account of Christianity requires those who would be Christian to lead lives of holiness in a manner that the church more often becomes a means to that end.

The Problem with Holiness

Part of the great genius of Catholicism was its ability to sustain Christianity as a way of life for peasants. To state the matter in this way will seem to many, both Catholic and Protestant, to put Catholicism in a negative light. Yet I mean such a characterization to be nothing but positive. First of all, I do not think there is anything wrong with being a "peasant," that is, someone who works every day at those duties necessary for us to eat, to have shelter, to sustain the having of children and to carry on the basic practices necessary to sustain communities. Peasants may not be "intellectuals," but they have knowledge habituated in their bodies that must be passed on from one generation to another. Peasants are often suspicious of intellectuals because they rightly worry about "ideas" that come from people who do not work with their hands.

Christian peasants usually do not think they are called to be holy. It is enough that they pray, obey and pay. Peasants, however, are usually more than willing to acknowledge the importance of holiness, venerating people, sacraments and relics that are clearly "different."[4] Peasants also know that those with authority over them, particularly those with ecclesiastical authority, may be anything but holy. Yet they understand the salvation offered by the church is not dependent on

---

[4]One of the remarkable themes of Caroline Walker Bynum's intriguing book *The Resurrection of the Body in Western Christianity, 200-1336* (New York: Columbia University Press, 1995) is how the "common piety" of "common Christians" acted as a check on the theological temptation toward a "spiritualization of the body." As she puts it, "Something very deep in third- and fourth-century assumptions was unwilling to jettison material continuity in return for philosophical consistency" (p. 68). At least part of that "something" was the veneration of relics of the saints whose bodies had already become relics through martyrdom and holiness while they were alive and so continue to be such after death. The Protestant rejection of such veneration as idolatry as well as the Enlightenment's characterization of such practices as superstition or magic most charitably can be described as superficial.

ecclesial representatives but rather is to be found in sacrament and saints.

No doubt "peasant Catholicism" was and is open to great perversions. Yet it is equally the case that one of the great virtues of such a Christianity is its capacity to be practiced by people who are poor. Such a Christianity is not a set of beliefs or doctrines one believes in order to be a Christian, but rather Christianity is to have one's body shaped, one's habits determined, in such a manner that the worship of God is unavoidable. Of course beliefs and doctrines matter, but it is not the peasant's task to insure they are rightly maintained. That is the task of the church, located in the office of the bishop, assisted by the theologian.

Whatever the limits and virtues this kind of Catholicism represented, it no longer seems to be an alternative for Protestants or Catholics who live in the so-called "developed societies." For in such societies no one any longer believes, in spite of evidence to the contrary, that they are peasants.[5] We believe that our lives are the outcome of choices we have made. Such a world seems, moreover, to be the kind of context that the pietist longed for. No longer is anyone made to be a Christian, but one only becomes a Christian through experience and voluntary commitment.

---

[5]Bankers, lawyers, university professors and other "professionals" have difficulty thinking of themselves as peasants. They certainly do not work with their hands. Yet in many ways they are much more at the mercy of those that rule than the traditional peasant. Perhaps their perilous state is the reason that they are so convinced that they are in control of their destinies, that is, that they are not peasants. Of course, that is why professionals are less free than peasants. Peasants know that there are masters and accordingly develop modes of resistance. Professionals think that they are the masters, which, of course, makes them completely incapable of defending themselves. Professionals present a peculiar challenge to the church because they assume that the church has no right to tell them what to believe or do. They assume that what they know as a banker or lawyer or university professor gives them a critical perspective on Christianity. Too often they think that they get to make up the kind of Christianity they like. It would be a step in the right direction if such people could see that, at least in terms of their ability to control their lives, they are not distinguishable from peasants. If such a recognition of their situation were possible, they might be able to develop peasant skills of resistance and survival—skills of hope that hopefully are drawn from practices provided by the church.

This view of the importance of individual commitment is often thought to be the great achievement of the Protestant Reformation. After all, is this not what the priesthood of believers is supposed to mean? The answer is, of course, "No"—if by the priesthood of believers you mean, as many American Christians seem to assume, that you can have an unmediated relation to God in which you are your own priest. Luther and Calvin would have been stunned by the suggestion that the church is a collection of individuals in which each person gets to determine his or her relation to God. Yet such individualistic presumptions have become the hallmark of Protestantism, particularly in America. Even more surprising, such individualism is underwritten by holiness traditions on the assumption that sanctification names our effort to lead a Christian life.

In such a context the emphasis on sanctification is not only Pelagian, but also often confuses middle-class moralism with "being Christian." Whatever the possibility Methodism may have represented for there to be a church that was at once evangelical and Catholic, it has been lost in the interest of being a church in which "anyone is welcome." Honesty compels me to acknowledge, of course, that there was never any great effort by American Methodists to embody Wesley's peculiar understanding of the church as a disciplined community. But whatever sense might have existed of that option is now submerged in debates about what techniques might provide for "church growth."

If we have any hope of reclaiming the church as a disciplined body of disciples, I believe some analogue to peasant Catholicism will be required. In other words, we need to recover the discipline of the body that at least offers an alternative to the endemic individualism and rationalism of modernity—an individualism and rationalism that has too often been thought necessary or at least confused with a concern with holiness. As a way to recover what holiness might look like as a discipline of the body I am going to discuss two quite different texts—Dale Martin's *The Corinthian Body* and Arthur Frank's *The Wounded Storyteller: Body, Illness, and Ethics*. Though these books treat quite different subject matters, I hope to show that they exhibit quite similar views of the body made possible by recent challenges to modernist notions of the self and, accordingly, can help us think better what a recovery by Christians of the sanctification of the body might look like.

The Permeable Body in Paul

Martin takes as his task nothing less than freeing Paul from our modernist reading habits exemplified by Cartesian soul-body dualism. Note the way I put the matter; that is, Martin does not claim he is telling us what Paul "really" meant by the "body," but rather he is trying to prepare us for being better readers of Paul by providing us with conceptual skills to make us better readers. Of course, he also thinks in the process we will better understand what Paul "meant," for example, in 1 Corinthians 12; but by that he means he expects us after having read his book to be able to make better sense of the many things Paul says about the body. In other words, Martin refuses to acknowledge a "text" that must be interpreted by an autonomous agent in the hope of discovering what the text "really meant."

The dualism between "text" and "interpreter" is but one form the body-soul dualism takes that his book is meant to challenge. Descartes gave the definitive expression to this dualism by associating body with matter, or nature, in an attempt to show the body can be explained in mechanistic terms. The soul, in contrast to the body, was identified as "mind," nonmatter, the supernatural and the spiritual.[6] Martin does not attribute the power and ubiquity of the body-soul dualism in modernity to Descartes, but rather Descartes provides for him one of the most compelling intellectual justifications of this dualism—a dualism Martin thinks disastrous for our understanding of Paul.

Drawing on ancient philosophical and, in particular, medical sources, Martin challenges our presumption that the ancients shared a Cartesian view of body-soul. Prior to modernity the soul, while certainly different from the body, was not conceived in the Cartesian sense as "nonmatter." Of course, there was a wide variety of ways the

---

[6]Dale Martin, *The Corinthian Body* (New Haven, Conn.: Yale University Press, 1995), p. 6.

soul and the body were understood in the ancient world, but Martin convincingly argues that however the soul was distinguished from body, the former was assumed to take up space, to consist of stuff. Accordingly our common distinctions between the inner and outer, or the individual and the social body, were unknown or understood quite differently. For example, it was assumed by many in the ancient world that the individual human body was but an instance of the social body.[7]

Martin does not provide this background in order to convince us that Paul was influenced by this or that ancient thinker but rather to help us understand that the various passages in Paul about the body, which initially may seem quite odd or at least unconnected, are not only interconnected but also make a great deal of sense. For example, Martin suggests that in 1 Corinthians 15:12-24 Paul meant that the resurrection of Christ necessarily entails the future resurrection of Christians. Paul assumes that Christian bodies have no integral individuality to them; so due to their existence "in Christ" through baptism they must experience the resurrection. "To deny the resurrection of their bodies is to deny the resurrection of Christ; to deny the resurrection of Christ is to render any future hope void. The Christian body has no meaning apart from its participation in the body of Christ."[8]

This perspective provides the background of Martin's main thesis that Paul, as well as most Corinthian Christians, saw the body as a dangerously permeable entity threatened by pollution. Against Paul there were at Corinth those who were economically better off, as well as of a higher social status, who stressed the hierarchal arrangement of the body without being overly concerned about the body's boundaries or pollution.[9] What is crucial for Martin's case is the recognition that the battle over hierarchy is a battle over the body. That some

---

[7]Ibid., p. 37.

[8]Ibid, p. 131. Martin rightly points out that this is why Paul sees no difficulty with the practice of baptism for the dead in 1 Corinthians 15:24-25. Such a suggestion makes for particularly embarrassing reading, given more orthodox Christian polemics against the Church of the Latter-Day Saints.

[9]Ibid., p. xv.

have become sick and even died as the result of their behavior at the Lord's Supper is what should be expected. Martin observes that Paul's advice to the better off at Corinth "flew in the face of the accepted practice and common sense of the upper class of Greco-Roman society. Paul's instructions represent a direct challenge to upper-class ideology and required that the higher-status Corinthians adjust both their expectations and their behavior to accommodate the needs of those of lower status."[10] This is not a surprising conclusion, but what Martin has helped us to see is that Paul assumes those upper class bodies are suffering because their bodies are not in fact "theirs."[11]

What this means is that our normal reading of 1 Corinthians 12:12-25 as a metaphor is a mistake. It is not as if the church is, like the body, interconnected, needing all its parts, even the inferior one. Rather, the church is the body from which we learn to understand our particular bodies. That is why Paul can argue that those who occupy positions of lower status are actually more essential and should be accorded more honor than those of higher status. As Martin observes, "This is not a compensatory move on Paul's part by means of which those of lower status are to be compensated for their low position by a benefaction of honor. Rather, his rhetoric pushes for an actual reversal of the normal, 'this worldly' attribution of honor and status. The lower is made higher, and the higher lower."[12]

When we read Paul, we best leave behind distinctions between the physiological and psychological we have come to think as obvious. Put colloquially, you have to get physical to read Paul. That is particularly true if we are to understand Paul's ethics and why his ethics cannot be distinguished from his theology. Paul presumes that the

---

[10]Ibid., pp. 74-75.

[11]Martin provides an illuminating account of the relation between class and the aesthetics of the body in the ancient world. People in the lower classes were assumed to be ugly, which of course meant that people in the upper classes policed their lives through medical practices meant to insure that they would be beautiful. Particularly important was how the infant body was treated, requiring quite detailed forms of swaddling (ibid., pp. 26-37). Against such a background one begins to appreciate how socially disruptive Paul's holding up the "weakest" member must have been.

[12]Ibid., p. 96.

church is Christ's body, so immorality is *not* like the body becoming ill or polluted; rather, it is to make the body ill and polluted. So questions regarding a man having sexual relations with his stepmother (1 Cor 5), Christian men using prostitutes (6:12-20), eating meat sacrificed to idols (8—10) and the proper eating of the Lord's Supper (11:17-34) are all connected. For Paul all of these matters are a question of the purity of the body and avoidance of pollution. For a Christian man to visit a prostitute is equivalent to the body being invaded by a disease that threatens everyone, since everyone is the body.

Crucial for Paul is the maintenance of boundaries of the body. Martin observes that just as Christian healers cured diseases by casting out demons, so Paul demands that the Corinthians cleanse the body by expelling those who pollute the church. Thus in 1 Corinthians 5:9-12 Paul tells the Corinthians they must not associate with the brother who is "sexually immoral or greedy or an idolater or a slanderer or a drunkard or rapacious," and in particular they are not to eat with such a man. Yet Paul says that they may have to associate with such people who are not in the church since otherwise Christians "would have to leave the cosmos entirely." Paul does not fear the pollution of the church by contact between a Christian and non-Christian, but he does think that the disguised presence within the church of a representative from the outside, from the cosmos, that should be "out there," threatens the whole body. The body of Christ is not polluted by mere contact with the cosmos or by the body's presence in the midst of the corrupt cosmos, but it may be polluted if its boundaries are permeated and an element of the cosmos gains entry into the body.[13]

The problem with modern readings of Paul, readings often meant to underwrite a concern with holiness, is that they are far too spiritual. For example, they assume that the problem with allowing someone who is sexually immoral to remain in the church is that he or she may tempt others to such immorality. Of course, that may be true, but that is not how Paul thinks about these matters. He thinks it has to do with the very constitution of the body. It is not a question of whether

[13]Ibid., p. 170.

sexual fidelity is or is not good for us. It is a question of the kind of body we are as the body of Christ. It is not a question whether we "understand" why we should marry or not marry. It is a question of how our bodies are positioned for the upbuilding of the body of Christ.[14]

Holiness is not, for Paul, a matter of individual will. Holiness is the result of our being made part of a body that makes it impossible for us to be anything other than disciples. That is why the "little things" matter. Whether one eats or does not eat meat sacrificed to idols matters because bodies do matter. How those "little things" shape and are shaped by the body may well change. Indeed, overconcentration on one "little thing" may well be a way to overlook or ignore what is really destructive for the holiness of the body. Yet what Martin has helped us to see is that our bodies—what we do and do not do, our habits—are what make our sanctification possible.

The Wounded Body

The great difficulty with Martin's account of Paul's letter to Corinth is not whether he gets Paul right or wrong but rather that we have no idea what to make of his account of the Pauline body. We can understand what he is saying, but we cannot imagine it in relation to our own lives. Few of us experience our bodies as he suggests the Corinthians experienced theirs. Our churches certainly do not think

---

[14]Martin rightly criticizes modern Protestant attempts to make Paul an advocate of marriage, family and heterosexual intercourse, e.g., why Paul really was not against marriage. According to Martin, even more misleading are attempts to make Paul a supporter of a "healthy view of human sexuality. As recent theorists of the history of sexuality have argued, the category itself—whether one is speaking of homosexuality, heterosexuality or human sexuality in general—is a modern one, heavily indebted to psychology, psychotherapy and the medicalization of the self so important to modern culture, especially since the nineteenth century. Paul does not speak of sexuality but of sexual actions and desires. And whenever the subject arises, Paul treats sex as potentially dangerous. If it cannot be completely avoided, it must be carefully controlled and regulated so as to avoid pollution and cosmic invasion" (ibid., p. 211). Martin's account of desire should be required reading for all who would develop an "ethics of sex" for today. Particularly important is Martin's contention that Paul shows no concern for the "propagation of the race" as the necessary presumption of Christian marriage (p. 214).

sinners should be expelled because they threaten the integrity of the body. After all, we join the church through voluntary commitment, by which we mean we have made an intellectual choice to be a member of this church. The body just comes along as part of the package.

In truth, we simply lack the resources, which is just another way to say examples, to imagine what Paul is talking about or how such a view of the body might fit over our own lives. Yet I think there is one place in our lives where Paul's understanding of the body is intensely exemplified—when we are sick. Sickness makes it impossible to avoid the reality of our bodies. When I am sick, I am not a mind with a suffering body; I am the suffering body. Illness may be the only time that we have the opportunity to discover that we are part of a story that we did not make up.

Arthur Frank observes that "seriously ill people are wounded not just in body but in voice."[15] The wound creates the need to tell stories because the diseased or injured body disrupts our old stories, creating the need for new stories. Yet these stories are not something separate from our bodies but rather come out of our bodies. Just to the extent, however, that stories give voice to our pain, and make our bodies familiar, the body eludes language. Desperate to domesticate the alienation our hurt creates between our stories and our bodies, we are driven to reduce the body to a thing, the topic of the story, the not me.[16] But in that reduction we lose our bodies as the source of who we are.

Another name for this reduction, according to Frank, is called modern medicine. Such a medicine is so close to us, having taught us how to understand ourselves, that we no longer sense its power over us. Frank calls our attention to the descriptive power of modern medicines by quoting a North African woman: "In the old days folk didn't know what illness was. They went to bed and they died. It's only nowadays that we've learned words like liver, lung, stomach, and I

[15]Arthur Frank, *The Wounded Storyteller: Body, Illness, and Ethics* (Chicago: University of Chicago Press, 1995), p. xii.
[16]Ibid., pp. 2-3.

don't know what!"[17] What it means for this woman (and for us) to seek medical care is not only to agree to follow the regimens prescribed but to tell her story in medical terms. To say how she feels, this North African woman is required to narrate her feelings in terms of a secondhand medical report. "The physician becomes the spokesperson for the disease, and the ill person's stories come to depend heavily on repetition of what the physician has said."[18]

Such a medicine is the institutional form of the Cartesian soul-body dualism Martin was combating in his attempt to exemplify Paul's understanding of the body. Something like the Cartesian dualism is required if the authority by which physicians impose their specialized language on their patients is to be legitimated. Frank characterizes this modern, medicalized body as monadic since it refuses to offer its pain to others or to receive reassurance from others that they recognize our affliction. A monadic body is the body modernist administrative systems desire since such bodies are all the more capable of being put through bureaucratic procedures. Moreover, it is hard to see how modern medical practice, which has increasingly become bureaucratic, could admit any other understanding of the body since the disease model that grounds its practice presupposes just such an account.[19]

The monadic body, however, is capable of permutations that Frank displays through ideal types he thinks exemplify modernist options: the disciplined and the mirroring body. The disciplined body is characterized by self-control, through which it seeks predictability by

---

[17]Ibid., pp. 4-5. Frank is quoting Pierre Bourdieu, *Outline for a Theory of Practice* (Cambridge: Cambridge University Press, 1977), p. 166.

[18]Frank, *Wounded Storyteller*, p. 6.

[19]Ibid., p. 36. Frank rightly observes that postmodernity is contradictory since it is so often characterized by opposing tendencies happening simultaneously. For example, "one side of postmodernity is the hyper-rationalization that subsumes the individuality extolled by modernity. Modernist medicine's general unifying view was a beneficent rationalization carried out in the interest of a science that had cure as it objective. DRGs [diagnostic related groups] are a less-than-beneficent rationalization carried out in the interest of cost-containment and administrative control over medicine. DRGs represent the modernist project turning against itself. A different side of postmodernity is the presence of self-stories that provide models of reclaiming the self" (pp. 70-71).

employing therapeutic regimens. Such regimens organize the body in the hopes that contingency can be avoided or at least compensated.[20] The mirroring body is at once its own instrument and object of consumption, that is, the body consumes and consumption enhances the body. Medicine, for example, is consumed in the interest of cure. This body is called mirroring because it recreates itself in the images of others. Like the disciplined body, the mirroring body fears contingency—not in the form of unpredictability but in the form of disfigurement. Both bodies are monadic, acting alone in a world that judges them, "but the judgments are made on different grounds: performance for the disciplined body and appearance for the mirroring body."[21]

Frank identifies another type, the dominating body, that is dyadic—not, however, in a sense of being for others but against others. For the characteristic feature of the dominating body is that it defines itself by force. When this body becomes ill, it gets mad, assuming the contingency of disease but never accepting it. The disciplined and mirroring bodies turn on themselves, but the dominating body turns on others.[22]

These types constitute modernist stories in miniature that shape and continue to shape our bodies. Frank reminds us that these are types, and thus that few of us ever exemplify any of them completely. Rather, we embody these types at different times in different circumstances, often several at the same time. Not to be forgotten, however, is that these are body types—that is, they remind us that just as there is no single thing called self, neither is there any single thing called body. Our bodies are under constant negotiation, shaped and reshaped by the stories in which we find ourselves.

Note, however, how these types, these stories that reflect the monadic body, also shape how we Christians now think of salvation and holiness. For illness, think of sin that we try to overcome through

---

[20]Ibid., p. 41.
[21]Ibid., p. 44.
[22]Ibid., pp. 48-49.

performance, appearance or, more likely, through dominion of others. Just as the monadic body in the name of independence becomes all the more subject to the power of medicine, so do we become what we fear just to the extent that we assume salvation is about making us safe from being dependent.

Yet Frank contends that these modernist stories that have legitimated and reinforced the power of medicine over our lives are now under challenge. He notes, "The *postmodern* experience of illness begins when ill people recognize that more is involved in their experience than the medical story can tell."[23] Correlative with this attempt to reclaim our stories of our illness according to Frank is a different body—the communicative body. It is not just an ideal type, but an idealized type that Frank commends exactly because it accepts its contingency as part of the fundamental contingency of life. If Frank is right about our discovery of such a body, it may moreover help Christians better understand what it means for our bodies to be made holy. One cannot help but hear theological resonance in Frank's account of the communicative body.

For example, he observes that the communicative body does not regard predictability as the rule but as the exception. Accordingly, there is no "self" desperately seeking to control the body. While there may be aspects of the body that are not the self, where one ends and the other begins is not easily determined. For the communicative body, the "body communes its story with others; the story invites others to recognize themselves in it. Thus the communicative body tells itself explicitly in stories. Reciprocally, stories are the medium of bodies seeking to approximate the communicative type."[24] For the communicative body, illness is not something that must be overcome, nor simply chaos, but rather an occasion for the discovery that we are on a journey, a quest, through which we learn "who I always have been" but did not know.[25]

The suggestion that the communicative body is not just an ideal

---

[23]Ibid., p. 6.
[24]Ibid., p. 50.
[25]Wendell Berry in *Another Turn of the Crank* (San Francisco: Counterpoint, 1996)

but an idealized type is Frank's way of saying that this is the way we ought to be. Moreover, he thinks that this is the way we can be "after modernity." It requires that ill people, not caregivers, need to regard themselves as the heroes of their own stories. This requires that we understand that heroism is not to be identified with those that can "do something" but rather is to be found in those who persevere through suffering.[26] What such a body offers is not victory but testimony. Our healing is not the overcoming of our illnesses but rather our ability to share our going on with one another through the community our stories create.

The imperative to receive testimony is, according to Frank, postmodern just to the extent that the witness must remain uncertain of what is received. Such uncertainty is required if we are not just to hear a story but to be implicated in it. As Frank puts it,

> This reciprocity of witnessing requires not one communicative body but a *relationship* of communicative bodies. Ordinary speech, conditioned by thinking on the model of law courts, refers to "the witness" as if witnessing could be a solitary act. Witnessing always implies a relationship; I tell myself stories all the time, but I cannot testify to myself alone. Part of what turns stories into testimony is the call made upon another person to receive that testimony. Testimony calls on its

---

notes that the metaphor of the body as a machine is accurate in some respects but must be controlled by "a sort of numinous intelligence." But in most ways the body is not like a machine: "the body is not formally self-sustained; its boundaries and outlines am not so exactly fixed. The body alone is not, property speaking, a body. Divided from its sources of air, food, drink, clothing, shelter, and companionship, a body is, property speaking, a cadaver, whereas a machine by itself, shut down or out of fuel, is still a machine. Merely as an organism the body lives and moves and has its being minute by minute, by an interinvolvement with other bodies and creatures, living and unliving, that is too complex to diagram or describe" (pp. 94-95). Berry argues in this wonderful chapter called "Health is Membership" that he would like to purge his mind and language of terms like *spiritual, physical, metaphysical* and *transcendental*—"all of which imply that the creation is divided into 'levels' that can be pulled apart and judged by human beings. I believe that the creation is one continuous fabric comprehending simultaneously what we mean by 'spirit' and what we mean by 'matter'" (pp. 90-91). Berry argues that the most important distinction is not between spirit and matter but between the organic and the mechanical.

[26]Frank, *Wounded Storyteller,* p. 134.

witnesses to become what none of us are yet, communicative bodies.[27]

Frank's account of the "wounded storyteller" is not located theologically or even in any specifiable tradition. In that way he remains ambiguously modern, that is, no longer confident of the modern self but yet having no alternative. He thinks we must reclaim from modern medicine the stories our bodies are trying to tell us, but the call for us to be the narrators of our own stories can be seen as but a new form of the modernist pretension that we can be our own creators. He quotes Paul Ricoeur's suggestion that each of us must become "the narrator of our own story without becoming the author of our life," noting that such a statement seems to require acceptance of divine authorship.[28] Yet, at least in this book, he offers us no sign that he believes in such an authorship.

He rightly observes that the ethic of testimony requires the witness to speak "outside the language of survival. Modernity disallows any language other than survival; the modernist hero cannot imagine any other way to be, which is why physicians are often genuinely baffled by criticism. People in post-modern times need different languages of meta-survival with various messages that death is all right. Clinical ethics needs these messages."[29] But "various messages" can be but an appeal to modem presumptions to let pluralism reign—we get to choose the afterlife we like.

I confess that I feel uneasy making these critical comments, given the testimony Frank has made through this book. I worry that my attempt to make Frank say where he stands is itself a modernist stance. Posing the question but reproduces the modernist assumption that what really matters is his "consciousness." What is important is that he is standing in stories that he cannot do without. For example, consider his own reflections on Paul:

> The body's story requires a character, but who the character is, is only created in the telling of the story. The character who is a communica-

---

[27]Ibid., p. 143.
[28]Ibid., p. 176.
[29]Ibid., p. 166.

> tive body must bear witness; witness requires voice as its medium, and voice finds its responsibility in witnessing. What is witnessed is memory, specifically embodied memory, a memory of experience written into the tissues. St. Paul, whose attitudes toward sexual embodiment are not popular, nevertheless expressed the embodiment of witness passionately. Paul knows he witnesses through his body: "In stripes, in imprisonments, in tumults, in labors, in watchings, in fastings" (2 Cor 6:5). Paul's ministry, to bring others into the body of Christ, is effected through rendering his own body available to suffering. This archetypal affinity of witness and bodily suffering cannot be evaded: Paul's unpopular message is that the responsibility of some is to find themselves called to the nexus of this affinity.[30]

So Paul's understanding of the body turns out to be the heart of Frank's perspective on illness. For Frank, "wounded storytellers" are moral witnesses, re-enchanting a disenchanted world. Through such stories we are reminded of our duties owed to the common sense world. The last line of Frank's book reads, "Illness stories provide glimpses of the perfection." Yet, is such perfection the perfection of Christ?

The Sanctified Body

I see no reason why we should try to give an answer in general to such a question. Rather, we should be grateful that people like Frank have been forced to challenge modernist presumptions about the self, renewing their and our imaginations in the process.[31] I purposefully use the language of "force" because I think that we do not develop new habits without being forced to do so. If we are to learn to think differently, we must have our bodies repositioned so that we have no

---

[30]Ibid., pp. 165-66.

[31]Frank confesses that his views on these matters are shaped by his belief that it is straightforwardly true, as we are told in the Gospel of John, that God loves the world. "I believe that the world was created and approved by love, that it subsists, coheres and endures by love, and that, insofar as it is redeemable, it can be redeemed only by love. I believe that divine love, incarnate and indwelling the world, summons the world always toward wholeness, which is reconciliation and atonement with God. I believe that health is wholeness" (ibid., p. 89).

other choice but to be what we were created to be.

This is the reason I have chosen to focus on the body as the subject of sanctification—and not just the body, but the sick body. By suggesting that Frank's account might help us to imagine what it would mean for us to be in the Pauline sense the body of Christ, I am seeking to find the means to remind us that perfection is but another name for submission. The focus on illness is anything but accidental. For illness is the means whereby we discover that we are subject to the world of necessity, that we are fundamentally dependent beings, that we need stories that shape our dying. Perfection is the art of dying. To practice that art requires that we learn the art of living as embodied members of Christ's body.

The problem with the language and practice of holiness in modernity is that it has been far too spiritual. To become holy has been presented as something we could become if we just tried hard enough. Using Frank's typology, holiness has been shaped by disciplinary, mirroring and dominating stories more than by the communicative story. We Christians have assumed modernist paradigms to shape our accounts of holiness because they promised to put us in control of our existence and, in particular, of our bodies. As a result our body's story, which is nothing less than the story of our desire for God, has been and is silenced.[32]

There are complex questions about the relation between the body and the body's stories that require further analysis.[33] I am sure that the way to understand how the body is storied is by looking more closely at how we are habituated. Indeed, I suspect that we are made perfect through our habits. So, for example, how we eat, with whom we eat,

---

[32]I have had to telescope Frank's quite wonderful discussion of the restitution and chaos illness narratives. The plot of the former is "I was healthy, I am sick, but tomorrow I will be healthy again." The chaos story appears to have no plot, seeing illness as the disruption of all stories. These stories can and are told in quite different ways, but Frank is surely right to see them as basic plot lines that shape our bodies not only when we are ill but when we are well. In an earlier book, *At the Will of the Body: Reflections on Illness* (Boston: Houghton Mifflin, 1991), Frank provides a wonderfully candid narrative of his own illness.

[33]As part of his analysis of the various types, Frank nicely shows what they imply about desire. For example, he notes that the disciplined body lacks desire, which but indi-

when we eat, what we eat, are among the most important questions of ethics. The monks have been right to think that nothing is more important for the shaping of communities of holiness than how our days are structured. Nothing is more important for holiness than learning to speak and the use of that speech to speak the truth to one another in love.

If we are to be so habituated, we will be so only by God's grace, which often is but another name for necessity.[34] Illness is one of the last sources of grace just to the extent that it forces us to need one another, to be communicative bodies, to be peasants. Perhaps that is why the most intense time in many Protestant worship services is when we share prayers for our own illnesses and those of others. Such times should not be depreciated, but they are not self-sustaining. For the body that we share with one another in prayer must become the body God shares with us in Eucharist. After all, the Eucharist is the only true sanctified body.

I was a member of Broadway United Methodist Church when I lived in South Bend, Indiana. We were located in a poor section of town. As part of our struggle to move to every-Sunday Eucharist we had discovered that we ought to provide a meal for the neighborhood after church. Most of the homeless who came to that meal did not come to church. One exception was an African-American woman. She was a classic bag lady.

As a former Evangelical United Brethren congregation we contin-

---

cates its impossibility (Frank, *Wounded Storyteller,* p. 41). The mirroring body produces desire, but what such a body wants is itself (p. 44). The communicative body desires and, so desiring, must learn to live out of control (p. 50).

[34]Some may be concerned that my account of the "self" may be too physical, risking some form of a reductionistic naturalism. These are complex questions in philosophical psychology about which I cannot pretend competency. It is my own view, however, that modern theological accounts of the self have not been sufficiently naturalistic. Theologians until recently have always known that humans are also creatures, which means we are animals. We should not be surprised, therefore, if modern science is capable of displaying our behavior in a predictive fashion. The mistake is to assume that this is sufficient reason to assume that all our behavior is "determined," whatever that would mean. For an account of these matters that nicely shows their complexity, see Owen Flanagan, *Self-Expressions: Mind, Morals, and the Meaning of Life* (New York: Oxford University Press, 1996).

ued to have spontaneous prayer requests during the church's prayer. We would pray for the end of war, to end hunger, for better race relations, but few would ever expose our hurts or fears. We would ask for Aunt Rose's illness to be healed, but we would not say if we were ill. That was not the case with the bag lady. She would say, "Lord, I am hurting, I have a cold, and I am frightened. Make these people help me."

It did not take long for her to teach us that we could so expose ourselves to one another. I believe that woman is a model for someone on the way to perfection. I believe that woman helped make us, Broadway United Methodist Church, Christ's body. That, I believe, is what the sanctified body looks like.

2

The Human Person as Intercessory Prayer

CRAIG KEEN

What follows was once two papers. They are now placed under a common title, barely separated, as if two subsections of the same continuous essay. They are not that, however. They remain very different; even as they are held together. The first is quite clearly a response to a position taken by Stanley Hauerwas. It was written to be spoken to those who had just heard Hauerwas's "keynote address." Its style is accessible by nonspecialists. It does not seriously engage the near or distant history of thought. It is in short a relatively light piece of nonacademic prose, though I hope not insubstantial. The second makes no direct reference to the work of Hauerwas. It was written to be heard, discussed, re*read,* and pondered by specialists. It is a continuous engagement with both the near and the distant history of thought. It is in short a heavy piece of academic prose. However, both pieces were written for the same time and place, are situated in the same set of questions, come to the same conclusion and could have been given the same title. I have considered rewriting them so that they might come together more peacefully; but it is perhaps better that they remain entangled as they were at first. If nothing else, let this be an experiment in intertextuality.

Perfect Agape[1]

Why perfection does not require a "self": an intriguing turn of phrase, especially for those children yet tracking campground mud across clean, middle class sanctuaries. Revival preachers indigenous to a different time and place once spoke unabashedly of such things. The way to be holy, the way to be perfect, they shouted (in no uncertain terms), is to "die-out to self." Even though this hyphenated word *die-out* remained overburdened with ambiguity, what a word it was! Had the evangelists who wrestled with us and the devil in those rings they called "tabernacles" not raised their voices, had they not spat out their hard consonants through sweaty faces and flailing arms, had they been more domesticated, this word would still have hit home and hit home hard, particularly where those already pummeled by identity crises lived. Some who contended there never made it out of the ring. Others who did still limp. I am not sure into which group I fall, but I think I fall into one. That is why I perk up—with dread and with hope—when I read Hauerwas's subtitle.

And there is much in his essay that is worthy at least of hope. Here one finds glimpses into an intriguing alternative to the modern *"self,"* that self most prominently delimited as Rene Descartes's famous dualism that pits soul against body. Hauerwas's focus is on a "permeable" bodily life that embraces and holds together everything that Descartes holds apart. This at least quasipostmodern view finds us humans to be a lively sociality. We are bodily beings who concretely interact at work and at play, in vital health and in incapacitating sickness, in the exuberance of life and on the pathway to death. Unless it walls itself

[1]To be read in English as much as in Greek.

off by isolating defense mechanisms, bodily life occurs, Hauerwas suggests, as genuinely communicative. The fragility of human beings in time issues an invitation to us to tell stories, to tell them and retell them; these stories are ever open to revision, because they are ever open to the unforeseen. Indeed truly humane stories must remain open, Hauerwas seems to say, because our lives together are finally at the mercy of a nearly naturalistic "necessity" beyond our control. Our lives change and often change drastically. We are kept always in an uncertain "contingency" that gives rise to the narratives that mark the moments of—what in narrative form has become—our "journey." But none of this is, for Hauerwas, the stuff either of individualistic intellectualism or individualistic pietism. There are no insulated minds or spirits here. Life is lived out as the bodily interchange of permeable human *lives.*

It is shod in these ideas (many of them entering the argument implicitly via Dale Martin and Arthur Frank) that Hauerwas takes a few steps into a doctrine of holiness. Here his words get particularly interesting.

Holiness is understood, of course, as a bodily phenomenon. And since "body" is not an "individuality" but a sociality, it is not surprising that the body held out to us here is "the body of Christ," i.e., the disciples of Christ, those who have (literally) entered into *his* permeable body. Holiness is "the work of the Holy Spirit building up the body of Christ" (p. 20). And how does the Holy Spirit do that? By making us "a disciplined body," a "habituated body," a submissive body, a body that practices "the art of living" and thus learns "the art of dying," a body so "shaped . . . that the worship of God is unavoidable" (p. 22). All of us human beings are the pattern of "what we do and do not do." We *are* "our habits." To enter into the body of Christ is to become *different,* because differently habituated. The real patterns of our real bodily lives change. Yet nobody changes in this way "without being forced to." It is here that "necessity" plays its part.

Life jerks us around. We think all is going well and suddenly we have the ground cut out from under us by illness—and by and by find ourselves perilously suspended above the abyss of death, helpless to do anything. Here—if we've not learned it before—we discover "that we are fundamentally dependent beings." Only thus out of control, led by God's grace ("which often is but another name for necessity"), do we together learn in the communicative reciprocity of bodily prayer (stories?) that life and death are a journey before and into the Eucharistic body of Christ. Thus we are sanctified. Jerked around—disciplined—by life together, we live with (and so tell stories to) one another and in word and deed are habituated precisely as our lives become a concrete journey that enters into the journey of Christ. The character thus habituated has a story-quality. It is a kind of history, but only because story is a kind of history.

Thus Hauerwas lays out the outlines of what we might not unfairly call his doctrine of holiness. Though no revival preacher would confuse what is said in his essay with the "dying-out to self" that figures so prominently in the holiness preaching of the misty past, what he says rings nonetheless true as an account of our lives together. Indeed there is a remarkable honesty to his words. So much so in fact that it seems that one is rather duty bound to pause before writing more. What Hauerwas says here should fill the lungs of those who have come to imagine their lives to be disembodied monadic solitudes. His words are fresh and clean and oxygen rich. Much that we are compelled to ask, even when we cannot articulate an intelligible question, he has answered. And yet I believe we may think—we may question—a bit more.

At rather important places in Hauerwas's essay—and this is most unexpected—there appear, of all things, intimations of naturalism. Indeed, the word itself makes its appearance, as does the related term *necessity*. Admittedly these words can be used in a wide variety of ways; but in what way do they contribute to what Hauerwas most wants to tell us? Do they not rather distract the reader from what is to be a postmodern reading of the human condition? Is it only a suspicious mind that disappointedly begins to wonder if perhaps *everything* that Hauerwas has said in this essay can be understood to be an account of the lives of folks who are *trapped* in an expanding universe, an expanding universe that makes itself felt not only at its cosmic edges but also in the hospital rooms and marketplaces of our little planet? It is very helpful, it seems to me, to think of human life as a bodily sociality. It is very helpful, it seems to me, to see that bodily sociality as a storytelling sociality. But what in that is elucidated by the notions of "necessity" and "naturalism"?

I am curious as to why movement into a holy life is described by Hauerwas as "forced." Of course, classic Protestantism, with which Wesleyanism is at least connected, knows that human beings are so ruined by sin that they cannot make even the first step toward God unless God's grace is at work. However, certainly in the Wesleyan tradition God's grace is never an overpowering force. God approaches lost sinners as a loving parent and from time to time leads a very concrete person in a very concrete way *from* death *into* a life of holiness. And even here where God is most intimately and lovingly laboring, there is the impossible possibility that one will not yield to God's grace but will resist the Spirit to the end. Had Hauerwas said that holiness is impossible without the miracle of prevenient grace, I would have no problem. And, of course, he could have used such words—in a certain way. However, grace is God, not a force. Is "the work of the Holy Spirit" understood in Hauerwas's essay to be very much like the blind necessity of a universe not quite at rest? I would think not. Yet why does it sound that way?

I am also uncertain about the way Hauerwas has chosen to connect the ironically healing effects of incapacitating illness with our realization that "we are fundamentally dependent beings" (p. 36). Admittedly, we are dependent beings. Certainly illness can make that as clear to us as anything. It seems also fair to suggest that illness can be instrumental in one's awakening to the mercy and holiness of God. However, long ago Rudolf Otto made clear that the business of holiness cannot be adequately elucidated by the notion of dependence.[2] Although Otto's thought has problems of its own, he is prob-

---

[2]Rudolf Otto, *The Idea of the Holy*, trans. John W. Harvey (New York: Oxford University Press, 1958), pp. 9-10.

ably more on track when he suggests that, before the holy, one is not dependent but, as Abraham says, "dust and ashes" (Gen 18:27). From such nothingness one does not learn that "perfection is but another name for submission." The hope of the cross (i.e., dust and ashes), the hope of the resurrection (i.e., the vitalizing work of the Holy Spirit), is that perfection is a liberation *from* deadly submission and *to* the newness of life.

Finally—and here I come closer to what is most dear to Hauerwas—I am not at all sure that a habituated life is a holy life, that "we are made perfect through our habits" (pp. 36-37). Of course, everything depends here on what is meant by perfection. I am not sure that Hauerwas has told us. But let us say that perfection from his point of view is hard to come by. Let us say that it is a life taught clear to the bone by hard days and hard nights. Let us say that it is a pattern of remaining open to the uncertainty of each new day and night. Let us say that perfection for Hauerwas is a life lived with others, that perfection has achieved a community solidarity by learning the pattern of telling significant others the story of our journey together. Let us say that perfection for Hauerwas is a life that in bodily sociality has learned the pattern of creatively weaving into its old story the new unforseen events that, when told, make that story no longer old but as fresh as the wound it helps heal. Let us say that perfection for Hauerwas is a guarded life, one that has been taught the pattern of drawing a clear line around the community that defines it, one that encourages community discipline and community character and community identity, one that knows the difference between us and them, one that is clean. Is this a desirable mode of life? Yes, no doubt often it is. Is this the kind of life that might be called "evangelical perfection"? I don't think so. For then we would find ourselves sent—and sent to the other, the outcast, the one not clean enough, the one defiled and defiling, the one without power, the one who will never be one of us, the one despised even by the noble peasant. In other words, although the perfection Hauerwas describes may well be perfect friendship, it is not perfect agape. Agape—and here I must ask one to listen hard to what must remain counterintuitive—opens wounds; it does not heal them. It opens the walls of communities, it does not guard them. It tells a story that even the most far-reaching and flexible narrative cannot get its arms around. It lives not for us but for those on the outside. It is not a perfection that is hard to come by. It is a gift, even if a rare gift. It is not taught by hard times but in spite of hard times, just as it is taught in spite of good times. It is an openness that prevails even when one can no longer cope with the chaos of another day, cannot say how the events of one's life are steps on a journey. Agape is perfection, holiness, because it is a kind of *ekstasis* that unravels every communitarian fabric, every story, every virtue, every habit.

Does this mean that *"community"* is to be jettisoned in some lonely return to individualistic pietism? Is there no *story* of the holy life? Does virtue, does habit, have no complicity with perfection? No. Not this. There are indeed a community and a story and a habituation that are hallowed. However, this community is ecclesial, *gathered*—and *gathered by* what can never be lodged *in* that community—gathered by what will only disruptively dwell there. And so the *story* of the community, however wordy it gets, however effectively it appropriates the events that befall it, must always come to silence—before an *ex*-propriating mystery that cannot be said. So too one's habits, as helpful as they are as a kind of collection of our worldly goods, are to be *offered*—in the freedom of the gift, the gift that is the Holy Spirit.

Thus "evangelical" perfection is a kenotic event that shifts the center of community, story, virtue and habit from itself to the glory of the coming God. It is thus before *this God's unsettling absence* that we cry out, "Woe is me!" and "Here am I; send me!" (Is 6:5, 8).

To Face Those Whom God Faces

The modern person. If it can be said that modernity has been a kind of presence moving on translucent wings, humming hypnotically about the enclosed boardrooms and battlefields, the accounting offices and lecture halls, the stock exchanges and laboratories of our time, then it can also be said that modernity began to emerge from its chrysalis during the early moments of the Renaissance and first spread its wings in Descartes. The word *modern* says "just now"[3] and its presence is perhaps best seen in its preoccupation with the highly abstract notion of the self, that identity of consciousness and of will, that judge of perceptions and of truth.

In Descartes the self is asserted with particular force only after a methodic doubt has laid to rest everything on which one might expect to be able to depend: authority, sense data, mathematics.[4] Indeed, the self breaks free precisely as it strives in mortal struggle against a hypothetical, but no less malignant, being who wields against all belief a terrifying omnipotence. Alone before the possibility of infinite delusion, flirting with madness, the solitary, momentary, faceless ego prevails: *I* stand as the single indubitable truth, the solid rock foundation of everything that might be judged to be true. Even if I were alone in the universe, even if every sight and sound were an illusion, even if 2 plus 3 did not equal 5, *I* would yet be. I so forcefully assert myself as the thinker even of the most wildly erroneous thought that it is impossible for this *I* not to be.[5]

---

[3]"Ad. Late L. *modern-us* (6th c.), f. *modo* just now" (*Oxford English Dictionary,* s.v. "modern").

[4]René Descartes, *Meditations on First Philosophy,* trans. Donald A. Cress (Indianapolis: Hackett, 1979), pp. 13-16.

[5]Ibid., p. 17:

> Therefore, am *I* not at least something? But *I* have already denied that *I* have any senses and any body. Still, *I* hesitate; for what follows from that? Am *I* so tied to the body and to the senses that *I* cannot exist without them? But *I* have persuaded *myself* that there is nothing at all in the world: no heaven, no earth, no minds, no bodies. Is it not then true that *I* do not exist? But certainly *I* should exist, if *I* were to persuade *myself* of something. But there is a deceiver (*I* know not who he is) powerful and sly in the highest degree, who is always purposely deceiving *me*. Then there is no doubt that *I* exist, if he deceives *me*. And deceive *me* as he will, he can never bring it about that *I* am nothing so long as *I* shall think that *I* am something. Thus it must be granted that, after weighing everything carefully and sufficiently, one must come to the considered judgment that the statement "*I* am, *I* exist" is necessarily true every time it is uttered by *me* or conceived in *my* mind (emphases added).

No single paragraph can adequately explicate what Descartes says of the "I am" in his *Meditations*. What is offered here is a perspective on Descartes's argument that is adjusted to what is to come in the work of later thinkers. However, Descartes, through

It is on Descartes's terms that the word *person* comes to be defined, e.g., by Leibniz as a monadic "self-consciousness and memory,"[6] by Locke as "a thinking intelligent being that can consider itself the same thinking thing in different times and places"[7] and most importantly by Kant as an "identity," a "permanence," a "substantiality of soul,"[8] as one whose rational, autonomous nature constitutes it as a singular "end in itself."[9] By and large a person has come to be one who is responsible for one's own actions, who executes one's own purposes, who is punished or rewarded for one's own deeds, who is thus the substantial owner of one's own dignity.[10]

Of course, not everyone has been entirely happy with the narrow, individual *scope* of Descartes's self-identical *cogito*. In fact, careful attempts have been made by very prominent modern thinkers to enlarge upon it so that it might include the manifold phenomena of sociality. Perhaps the most impressive and ambitious of these is Hegel's. Hegel's world is conflictual, one in which opposing entities, clashing, vying with one another for supremacy, make their way out of a virtual chaos. The person, the singular self, in Hegel is that unity that without such conflict remains an empty abstraction but with it is enriched and matured.[11] Deep within every sleeping person is folded the potency of a spirit that has embraced its world and made it a part

---

the opening of his second meditation, is indeed a foundation builder, and the foundation he builds is an "I" that is so unrestrictedly asserted that not even an omnipotent being can shake it. Even knowing it as certain (the explicit point of the inquiry) is finally secondary to its being asserted.

[6]Gottfried Leibniz, "On the Active Force of Body, On the Soul and on the Soul of Brutes (Letter to Wagner, 1710)," trans. George Martin Duncan, in *Leibniz: Selections*, ed. Philip P. Wiener (New York: Scribner's, 1951), p. 507.

[7]John Locke, "An Essay Concerning Human Understanding," *The English Philosophers from Bacon to Mill*, ed. Edwin A. Burtt (New York: Modern Library, 1967), p. 315.

[8]Immanuel Kant, *Critique of Pure Reason*, trans. Norman Kemp Smith (New York: Macmillan, 1958), pp. 341-44.

[9]Immanuel Kant, *Grounding for the Metaphysics of Morals*, trans. James W. Ellington (Indianapolis: Hackett, 1981), p. 36.

[10]Arthur C. Danto, "Persons," in *The Encyclopedia of Philosophy*, ed. Paul Edwards (New York: Macmillan, 1967), 6:110-13; and John H. Lavely, "Personalism," in *Encyclopedia of Philosophy*, 6:107-9.

[11]G. W. F. Hegel, *Phenomenology of Spirit*, trans. A. V. Miller (New York: Oxford University Press, 1977), pp. 113-14, 290-91.

of herself. Indeed, the fully self-actualized person is one who by strife has found herself displaced in others and has won herself back again by appropriating them, by reconciling them to herself.[12] It is the person lost to find, gone to battle to return home, decentered to center, that becomes a self-identical unity once more. The *meaning* of "person" according to Hegel is enriched self-identity. In fact, the whole of universal history is finally for him the absolute return of the most extreme investment, an investment in which the Absolute Spirit, the ultimate self, sacrifices itself, consumes itself, that it might rise like a phoenix but with greater freedom and strength and glory.[13] In the end Odysseus comes home more substantially and independently Odysseus than he would ever have been without the sacred and profane conquests of his odyssey.

Coming from the outside. The ultimately substantial and independent subject of Descartes and Hegel is not as unquestionable as it appears, however. Despite claims to the contrary, the famous phrase *"cogito ergo sum"* and the more direct "I am" remain quite indeterminate. What is it that gives to these phrases their purported immediate certainty? Has "intuition" really shaken free from custom and habit and "got hold of its object purely and nakedly as 'the thing itself'?"[14] Indeed, there seems to be nothing to indicate that Descartes's formulations come to rest in and for themselves either as the simple unity at the heart of his *Meditations* or as the complex unity at the heart of Hegel's *Phenomenology*. Rather, each moves as a tangle of strands of significance that trail off in every direction. Thus Nietzsche finds in the little term *cogito* an insuperable and insubstantial complexity that Descartes's meditations bypass in their haste to abide by "our grammatical custom that adds a doer to every deed."[15]

If the "I think" is not a singularity, then Descartes's solid rock foundation shatters and with it the modern spirit. An "I" become *essentially*

---

[12]Ibid., pp. 356-58, 384-85.
[13]Ibid., pp. 11-12, 14.
[14]Friedrich Nietzsche, *Beyond Good and Evil*, trans. Walter Kaufmann (New York: Vintage-Random, 1966), p. 23.
[15]Friedrich Nietzsche, *The Will to Power*, trans. Walter Kaufmann (New York: Vintage-Random, 1967), p. 268.

complex is an "I" that has lost control; and an "I" that has lost control is no longer a modern person. Could it then be that there is no self at the center of its world, no substance or subject that in the face of disaster would remain or once again become self-identical, essentially the same, the inviolable owner of its properties? Perhaps it is rather the case that, as Nietzsche stated, "the properties of a thing are effects of other 'things,'" that "if one removes other 'things,' then a thing has no properties," i.e., perhaps "there is no thing without other things." Perhaps "if I remove all relationships . . . of a thing, the thing does not remain over." Perhaps the whole world "is essentially a world of relationships" and thus a world with shifting centers, a world with no totalizing unity.[16] If so, then "the assumption of one single subject is perhaps unnecessary; perhaps it is just as permissible to assume a multiplicity of subjects, whose interaction and struggle is the basis of our thought and our consciousness in general." In that case one might hypothesize: not the subject as a unity, but the "subject as multiplicity," a nodal point of convergence of the lines of a network of relations, a motion of inevitable slippage to those others in whom it is intimately involved.[17]

Consider this: Let us say that a young woman has resolved to give something to a poor beggar on the street. However, she wants to give

---

[16]Ibid., pp. 302, 306.

[17]Ibid., pp. 270-71. Nietzsche's account of the relationality of the world weaves into its odd metaphysical aphorisms the deceptive phrase "the will to power." The phrase is perhaps best understood not as "the will to acquire power," but as "the will to expend power," e.g., artistically. However, there is much in Nietzsche that does not escape the grasp of the age he so frequently and so vigorously criticizes. For example, Nietzsche's metaphysical use of the notion of the "eternal recurrence of the same" may well be a survival of modern totalization. Although this notion does not function metaphysically (and may indeed function in a non-totalizing manner) in *Thus Spoke Zarathustra* (trans. Walter Kaufmann [New York: Viking, 1954], see e.g., pp. 157-60, 215-21), it does at least at times in *Will to Power* (e.g., pp. 544-50). Nietzsche's writings get loosed from their modern metaphysical entanglements (while remaining no less entangled) in the philosophical discourse of the late twentieth century, and in particular in the tradition of Martin Heidegger. Although what occurs late in the century is already to some degree taking place in Heidegger's *Being and Time* (trans. John Macquarrie and Edward Robinson [New York: Harper, 1927]), it does not do so openly until Heidegger is read in earnest as an outsider in relation to "existentialism." The same may also be able to be said of the writings of Søren Kierkegaard.

this time with an uncommon purity. This time she wants no adulation, no appreciation, no prospect of a return on an investment. She wants simply to give, as if her right hand did not know what her left hand were doing, as if she had lost herself, as if she were not in control. The more she calculates toward giving in this way, the more giving eludes her.[18] A gift *qua* gift unsettles a substantial subject. In the words of the French philosopher Jacques Derrida, "One would even be tempted to say that [such] a subject as such never gives or receives a gift."[19] As long as one stands as the identity who gives, no gift occurs. As long as the other stands as the identity who receives, no gift occurs. Rather, a transaction takes place. The gift becomes property.

Thus when a gift is truly given, giver and receiver are thrust back from one another just as they are held together. They are bathed in the *gift's* light, become *its* dependents.[20] The giver and receiver are not two independent entities who touch only via an item that passes from one right hand to another. There is rather something in the space *between* these two that prevents them from existing in themselves and for themselves, from taking control of the situation. There is something here that grants the fluctuating movement in which you and I meet, in which I hear your alien call. I am gifted by your call—by the otherness that sustains you, an otherness that gathers me and turns me to you.

---

[18]"The moment the gift, however generous it be, is infected with the slightest hint of calculation, the moment it takes account of knowledge *[connaissance]* or recognition *[reconnaissance]*, it falls within the ambit of an economy: it exchanges, in short it gives counterfeit money, since it gives in exchange for payment. Even if it gives 'true' money, the alteration of the gift into a form of calculation immediately destroys the value of the very thing that is given; it destroys it as if from the inside. The money may keep its value but it is no longer given as such. Once it is tied to remuneration *(merces)*, it is counterfeit because it is mercenary and mercantile; even if it is real" (Jacques Derrida, *The Gift of Death*, trans. David Wills [Chicago: University of Chicago Press, 1995], p. 112).

[19]Jacques Derrida, *Given Time 1: Counterfeit Money*, trans. Peggy Kamuf (Chicago: University of Chicago Press, 1992), p. 24.

[20]Ibid., pp. 24, 40. See Martin Heidegger, "The Origin of the Work of Art," trans. Albert Hofstadler, in *Martin Heidegger: Basic Writings*, ed. David Ferrell Krell (San Francisco: Harper & Row, 1993), p. 143: "As necessary as the artist is the origin of the work in a different way than the work is the origin of the artist, so it is equally certain that, in a still different way, art is the origin of both artist and work."

It is the relationality of the gift that makes the giver everything she is by bringing her to a troubling acknowledgment of what she is not and never can be. She is a "who" not in herself but in answer to the other's call. I am no personal identity. I am only *because* an other calls out to me and elicits the reply, "Here I am." The call of the other indeed "somehow precedes . . . [the giver's] identification with itself, for to this call I can *only* answer, have already answered, even if I think I am answering 'no.'"[21]

God vis-à-vis. Had the word *person* not accepted the feast-masters' invitation to occupy the honorific seat "self as dignitary" at its celebration, the critiques offered by Nietzsche and Derrida would perhaps not have threatened it. Indeed "person" might have gone so far as to take its place in the ranks of their corps against the whole modern

---

[21]Jacques Derrida, "'Eating Well,' or the Calculation of the Subject," in *Points . . . : Interviews, 1974-1994,* ed. Elisabeth Weber, trans. Peggy Kamuf et al. (Stanford, Calif.: Stanford University Press, 1995), p. 261. The relationality of giver and other is the relationality of language, for Derrida. Language is the network of relations, the openness that grants the nonsubstantial "who" called forth in every noneconomic gift. Relationality has priority over those related. Therefore, language has priority over those who speak and write.

> Now if we refer, once again, to semiological difference, of what does [Ferdinand] Saussure, in particular, remind us? That "language [which only consists of differences] is not a function of the speaking subject." This implies that the subject (in its identity with itself, or eventually in its consciousness of its identity with itself, its self-consciousness) is inscribed in language, is a "function" of language, becomes a *speaking* subject only by making its speech conform—even in so-called "creation," or in so-called "transgression"—to the system of the rules of language as a system of differences, or at very least by conforming to the general law of *différance,* or by adhering to the principle of language which Saussure says is "spoken language minus speech." . . . It is the domination of beings that *différance* everywhere comes to solicit, in the sense that *sollicitare,* in old Latin, means to shake as a whole, to make tremble in entirety. Therefore, it is the determination of Being as presence or as beingness that is interrogated by the thought of *différance.* . . . First consequence: *différance* is not a present being, however excellent, unique, principal, or transcendent. . . . This unnameable is the play which makes possible nominal effects . . . that are called names, the chains of substitutions of names in which for example, the nominal effect *différance* is itself *enmeshed,* carried off, reinscribed, just as a false entry or a false exit is still part of the game, a function of the system. (Jacques Derrida, "*Différance,*" trans. Alan Bass, in *Deconstruction in Context,* ed. Mark C. Taylor [Chicago: University of Chicago Press, 1986], pp. 408, 409, 413-14, 419)

program. However, this was not to be. Modern thinkers saw in the term something that they could exploit, and their invitation was too much to be refused. Of course, there was much to the person that would not fit at all. But the modern mind is nothing if not resourceful, and a term that might have been utterly unmarketable became profitable indeed. Thus it was that in the Age of Reason—an age in which knowledge is the power to harness untapped resources, to have and hold and be, to conquer, to identify, to reward and to punish—in this age "person" became a modern man.

Admittedly, already in the early sixth century Boethius had defined *person* in a way that recommended it to the children of Descartes. "A person," he said, "is an individual substance of a rational nature." However, this definition makes a significant turn from the trajectory of the word as one finds it in earlier (and much of later) trinitarian discourse.

The Latin *persona* ("person") becomes a serious theological term for the first time in the writings of Tertullian. The word seems at first to have been equivalent to the English *mask* or *face,* but also to a whole host of relations: those of the family, those that constitute friends or enemies, those at work in discourse. It is used by the Stoic philosopher Epictetus to indicate the modes of life given by providence: one is wise who lives one's *persona* as the divine playwright has written it. Even in Tertullian *persona* is not yet a technically precise theological term. He uses it often loosely, and he uses it in many ways, for example, in its older sense of "'mask,' . . . [or] 'face,'" he uses it "in a quasi-dramatic sense, [as well as] in a sense equivalent to *homo* (human) or *vir* (male adult)."[22] At the same time Tertullian was making theological use of *persona,* his contemporary, Hippolytus, was similarly using the Greek *prosōpon.*[23] *Prosōpon* seems also (and even more primarily) to have meant "face."[24] However, in Hippolytus it comes to signify, as *persona* does in Tertullian, the distinctiveness

[22]Edward J. Fortman, *The Triune God: A Historical Study of the Doctrine of the Trinity* (Grand Rapids, Mich.: Baker, 1972), p. 113.

[23]Ibid. See also J. N. D. Kelly, *Early Christian Doctrines* (San Francisco: Harper & Row, 1978), 110-15.

[24]"F. *pros* to + *ops, op-* eye, face" (*Oxford English Dictionary, s.v.* "prosopalgia").

of the Son of God and of the Holy Spirit in relation to each other and in relation to God the Father.[25] Here the Greek and Latin terms begin to take on a profound significance. The words speak—as does the English *face*—of a distinctiveness, a uniqueness, which is simultaneously a relatedness. The Father and the Son and the Spirit are thought here as different, certainly. However, they are not different from each other in the way Aristotelian substances are.[26] These are *three* only as each faces the others and is faced by them. However, they are also *one* as they face each other.

Tertullian invented the word *trinitas* to get at the complexity and oddity of the notion that what is three *personae,* three *prosōpa* is also one. However, in order specifically to say "one," Tertullian chose the Latin *substantia.*[27] The Trinity is one *substantia,* one solidarity, *as* it is three *personae,* three faces.

Among Greek-speaking theologians in the patristic era a similar move is made. The Greek term *ousia* has been not uncommonly taken as equivalent to the Latin *substantia.* It too signifies concrete reality.[28] Therefore, it was not overly difficult for Greek theologians to acknowledge that the trinitarian God is one *ousia.* Their difficulty came as they tried to find the words to say what they very much

---

[25]In neither Hippolytus nor Tertullian, however, did the term involve "the idea of self-consciousness nowadays associated with 'person' and 'personal'" (Kelly, *Early Christian Doctrines,* p. 115).

[26]Consider the following passage from Tertullian. He has just been discussing the Son's difference from and identity with the Father. He compares that difference and identity to a similar relation that obtains between oneself and one's thought: "Whatever you think, there is a word; whatever you conceive, there is reason. You must needs speak it in your mind; and while you are speaking, you admit speech as an interlocutor with you, involved in which there is this very reason, whereby, while in thought you are holding converse with your word, you are (by reciprocal action) producing thought by means of that converse with your word. Thus, in a certain sense, the word is a second *person* within you, through which in thinking you utter speech, and through which also, (by reciprocity of process,) in uttering speech you generate thought. The word is itself a different thing from yourself" (Tertullian *Against Praxeas* 5.19, trans. Peter Holmes, in *The Ante-Nicene Fathers,* ed. Alexander Roberts and James Donaldson, vol. 3, *Latin Christianity: Its Founder, Tertullian* (Buffalo, N.Y.: Christian Literature, 1885), p. 601.

[27]Kelly, *Early Christian Doctrines,* p. 114.

[28]Ibid., p. 129.

believed: that the Trinity is one because God is three. Although Hippolytus had already addressed this problem by his use of *prosōpon,* the mainstream of Greek theology found that word too weak, at least alone. The word was not simply rejected; indeed one finds it in the Creed of Chalcedon. However, another moved alongside of it, then took its place. In order to make clear that the three—Father, Son, and Spirit—are not mere labels that we have ignorantly pasted on the one homogeneous, monarchical God but are genuinely real, they turned to the word *hypostasis.* God, the holy Trinity, they taught, is one *ousia* and three *hypostaseis.*

The irony is that *hypostasis* and *ousia* "were originally synonyms."[29] In fact, *hypostasis* is a more natural equivalent of *substantia.* Nevertheless, the Greek formula distinguishes between the terms. This is not a denial of the relationality of the three; it is, however, an affirmation of their reality. Whereas the Latin says emphatically that the Father and the Son and the Spirit are the *faces* of God, the Greek says emphatically that those faces are *real.*[30]

The radical nature of this idea may be seen to be at work in another Greek word that came to occupy the contentious center of the Arian controversy of the fourth century. The word is *homoousios.* What this word said to the defenders of the Creed of Nicea is that the incarnate one is truly and unequivocally God. What it is for Jesus Christ to be God is identical to what it is for the eternal Father to be God.[31] With the later declaration in the Creed of Constantinople that the Holy Spirit is "worshiped and glorified together with the Father and the Son," the significance of the term widened. It comes now to be understood that whatever makes any one of the three persons God is identical to what makes any one of the others God. Indeed, to approach any one of the three is to approach the entirety of the Godhead.[32] The Father remains the wellspring of deity. The Son is deity

---

[29]Ibid.

[30]Cf. John Meyendorff, *Christ in Eastern Christian Thought* (Crestwood, N.Y.: St. Vladimir's Seminary Press, 1987), p. 66.

[31]See, for example, Kelly, *Early Christian Doctrines,* p. 245.

[32]Augustine, *The Trinity,* trans. Stephen McKenna (Washington, D.C.: Catholic University of America Press, 1963), pp. 243-44.

"generated" or "begotten." The Spirit is deity "proceeding" or "spirating." However, they are each fully God, and there is no fourth, no generic deity, standing beyond these three as that of which they are expressions. There is no God but the one in three and three in one.

Further, this one is three and these three are one as the Father, the Son and the Holy Spirit are mutually dependent. Although the Father is as such the source of deity, he is so only as the Son is generated and as the Spirit proceeds. Not only is there no Son or Spirit without the Father, there is no Father without the Son and the Spirit, as Gregory of Nyssa insisted.[33] Though not to be confused with one another, they are also not to be ripped apart from one another.[34] Indeed, each is the entirety of deity precisely because it cannot be ripped apart from the others. To deal with the Son is always also to deal with the Father and the Spirit. One person inevitably slips to the others in whom it is intimately involved.

The same dynamics of interdependence among the Trinitarian Persons is affirmed by theologians of the West. It is Augustine who is decisive here.[35] Although his starting point, unlike that of the Eastern

---

[33]Gregory of Nyssa, *Against Eunomius*, trans. H. C. Ogle, rev. H. A. Wilson, in *A Select Library of Nicene and Post-Nicene Fathers of the Christian Church*, second series, vol. 5, *Gregory of Nyssa: Dogmatic Treatises, etc.* (Grand Rapids, Mich.: Eerdmans, 1954), p. 102.

[34]"Basil remarks, 'Everything that the Father is is seen in the Son, and everything that the Son is belongs to the Father. *The Son in His entirety abides in the Father*, and in return *possesses the Father in entirety in Himself*. Thus the hypostasis of the Son is, so to speak, the *form and presentation* by which the Father is known, and the Father's hypostasis is recognized in the form of the Son.' Here we have the doctrine of the co-inherence, or as it was later called 'perichoresis,' of the divine Persons. . . . [The] distinction of the Persons is grounded in Their origin and *mutual relations*. They are, we should observe, so many *ways* in which the one indivisible divine substance *distributes and presents Itself*, and hence They come to be termed 'modes of coming to be' (*tropoi hyparxeos*). So Basil's friend Amphilochius of Iconium . . . suggests that the names Father, Son and Holy Spirit do not stand for essence or being ('God' does), but for *'a mode of existence or relation' (tropos hyparxeos etoun scheseos)*. . . . None of the Persons possesses a separate operation of His own, but one identical energy passes through all Three" (Kelly, *Early Christian Doctrines*, pp. 264, 265-66, 267 [emphasis added]).

[35]But see also Kelly's account of the position of Victorinus: "God is essentially in motion, and in fact His *esse* is equivalent to *moveri*. . . . Again and again he insists on the circumincession, or mutual indwelling, of the Persons (e.g., *omnes in alternis existentes*)" (Kelly, *Early Christian Doctrines*, pp. 270, 271).

Cappadocians, is the divine *substantia* rather than the divine *personae,* Augustine is (according to J. N. D. Kelly) nonetheless convinced that the unity of God is to be found only as the *personae* "severally indwell or coinhere with each other," just as "the distinction of the Persons . . . is grounded in Their mutual relations within the Godhead."[36] He clearly struggles with the idea. He concludes at last, in congruity on this point with the East, that a Trinitarian Person *is* a real relation. Kelly writes:

> His own positive theory was the original and, for the history of Western Trinitarianism, highly important one that the Three are real or subsistent relations. . . . Father, Son and Spirit are thus relations in the sense that whatever each of Them is, He is in relation to one or both of the others.[37]

In trinitarian discourse a person is thus not a self-identical subject; it is not a substance that *has* relations. Rather it *is* as such a relation, it is a reality *only as* a relation. A person is a face that is what it is only as it meets other faces. The Trinity is one concrete reality only as this *vis-à-vis.*[38] There is nothing, not even a divine reality, standing behind the faces that are the Trinity. Divine reality is the relationality of the faces that are the Father, the Son and the Holy Spirit.

God facing the world. Following the solidification of the orthodox doctrine of the Trinity, attention came to be given more and more to the problem of the relation between the divine and human natures of Christ. As it became increasingly clear that Christ is both *fully* human and *fully* divine, and as language for the relation between those natures became more and more crucial, the terms that played so great a part in the history of the doctrine of the Trinity made themselves available once more.

---

[36]Kelly, *Early Christian Doctrines,* pp. 273-74. In Augustine's own words: "Although, to tell the truth, it is difficult to see how one can speak of the Father alone or the Son alone, since the Father is with the Son, and the Son with the Father always and inseparably, not that both are the Father or both the Son, but because they are always mutually in one another and neither is alone" (Augustine, *Trinity,* p. 209).

[37]Kelly, *Early Christian Doctrines,* pp. 274-75.

[38]"We do not call these three together one person" (Augustine, *Trinity,* p. 236).

The Creed of Chalcedon affirms that a solidly human reality and a solidly divine reality converge in the one Jesus Christ.[39] Yet what makes him one is the radical concurrence of his humanity (*homoousion hemin* [*homoousios* with us]) with his deity (*homoousion to patri* [*homoousios* with the Father]). These two natures are one precisely *in* the *prosōpon*, the *hypostasis*, of the divine Word, the person of the outgoing address of God. Human being is concretely here *as* a *divine* relationality. Leontius of Jerusalem in the sixth century is apparently the first to describe this concurrence as "enhypostatic," that is, that Jesus Christ is a human person only as he gives his life and destiny utterly in reply to the person of the divine Word.[40] He is a human person only as the concreteness of his undiluted human being moves *into* the outgoing movement of the second person of the Trinity.[41] That is, *person* is still in the first place the manner in which *God* moves, the way *God* faces. But the doctrine of the enhypostaton makes the human a person as well. In the incarnate second Adam the human is destined through and through to be a person, to face God without remainder and in so doing to face what God faces in God's outgoing, to turn without remainder to those to whom God's face is turned.

Karl Barth

Perhaps the most influential treatment in our century both of the doctrine of the Trinity and of the person of Christ is to be found in the theology of Karl Barth. Barth's is a theology of revelation. It is cer-

[39]Meyendorff, *Christ in Eastern Christian Thought*, pp. 29-46. The text of the creed may be found in "Definition of Chalcedon," trans. Albert C. Outler, in *Creeds of the Churches*, ed. John H. Leith (Atlanta: John Knox, 1982), p. 36.

[40]Meyendorff, *Christ in Eastern Christian Thought*, pp. 73-79. See also pp. 61-68.

[41]"The hypostasis is not the product of nature: it is that in which nature exists, the very principle of its existence. Such a conception of hypostasis can be applied to Christology, since it implies the existence of a fully human existence, without any limitation, 'enhypostatized' in the Word, who is a divine hypostasis. This conception assumes that *God, as personal being, is not totally bound to his own nature; the hypostatic existence is flexible, 'open'*; it admits the possibility of divine acts outside of the nature (energies) and implies that God can personally and freely assume a fully human existence while remaining God, whose nature remains completely transcendent" (Meyendorff, *Christ in Eastern Christian Thought*, p. 77 [emphasis added]).

tainly also a theology of the wholly other; but Barth speaks of this God only as he speaks of the apocalyptic event in which God is made known, in which God dramatically opens to what God is not. Indeed, everything Barth says emerges finally from that event in which two utterly alien realities—one creature, the other Creator—become unconfusedly but also undividedly one.[42]

It is because of the exhaustively constitutive nature of God's self-revelation in Barth's theology that late in his career he could write of "the humanity of God."[43] The wholly other comes so close that one cannot speak of God in isolation from those to whom God comes. The place of God's coming is the history of the concrete human being, Jesus Christ. The unity of this history occurs as a movement of mutual self-giving, of mutual kenotic love.[44] It is the "yes" spoken by each to the other; the concrete and particular event of the absolute openness of the human Jesus to the God who is absolutely open to him. The history of Jesus Christ is the history in which there is no distinguishing what human being is about from what God is about.[45] Jesus Christ is human being corresponding *(entsprechend)* to God.[46] The history of Jesus reveals that both the truth of God and the truth of human being is an outgoing movement of self-giving. They are not *entities* that go out, that love; they *are* the event of love.[47]

---

[42]"Now the absoluteness of God strictly understood in this sense means that God has the freedom to be present with that which is not God, to communicate Himself and unite Himself with the other and the other with Himself, in a way which utterly surpasses all that can be effected in regard to reciprocal presence, communion and fellowship between other beings" (Karl Barth, *Church Dogmatics,* vol. II, *The Doctrine of God,* trans. T. H. L. Parker et al., ed. G. W. Bromiley and T. F. Torrance (Edinburgh: T & T Clark, 1957), 1:313.

[43]Barth, *Church Dogmatics,* vol. III, *The Doctrine of Creation,* trans. T. H. L. Parker et al., ed. G. W. Bromiley and T. F. Torrance (Edinburgh: T & T Clark, 1958-1961), 2:62; vol. IV, *The Doctrine of Reconciliation,* trans. G. W. Bromiley, ed. G. W. Bromiley and T. F. Torrance (Edinburgh: T & T Clark, 1956-1969), 2:72, 519; and *The Humanity of God,* trans. John Newton Thomas and Thomas Wieser (Atlanta: John Knox Press, 1960), p. 46.

[44]Barth, *Church Dogmatics* IV/1:§59.1; IV/2:§64.

[45]Ibid., III/2:61-74, 161-77.

[46]Ibid., III/1:197.

[47]Ibid., IV/2:§68. "It is obviously true of man in general that his being is not to be sought behind or apart from . . . [historical] movement, as if it were first something in itself which is then caught up in this movement. . . . Whatever his state may be, he is

That God's self-revelation as the history of Jesus is for us, that it reaches out and gathers us to itself as our truth, is the work of the Holy Spirit, without whom nothing of God is revealed. The Father is the mystery that opens to human being (*der Offenbarer* [the revealer]), the Son is that openness as it penetrates the entirety of human being (*die Offenbarung* [the revelation]), but it is the Spirit that is the embrace of openness that gathers human being to itself (*die Offenbarsein* [the act of being revelatory]).[48] Spirit, Son, Father: none of these *is* without the others. Thus for Barth there is in God nothing that is static. God is the occurrent Trinity through and through, not a substance that occurs.[49] Thus, in Christ the center of human being is outside itself in the outgoing trinitarian movement of the Father to the Son by the Spirit, a movement that is simultaneously in the Spirit, through the Son, to the Father. To look to human being is to look to the shifting centers of the Trinity that open that being beyond itself.[50]

Further, the outgoing love of God occurs as the Son travels into "the far country," the region where human being plunges into the abyss of sin and death.[51] God's outgoing to the godless is thus God's self-humiliation, a humiliation that is never simply left behind.[52] As the Son returns home, exalting human being "in the closest proximity to God from the greatest distance," he is still the crucified one.[53] To be human is therefore to move down the path of the history of Jesus Christ, in the rhythm of the crucifixion and the resurrection, turned to the God who is love and with this God to the lost and abandoned.

---

only *in* this state; it is not his being but only the attribute and modality of his being. His true being is his being in history grounded in the man Jesus, in which God wills to be for him and he may be for God" (*Church Dogmatics* III/2:162).

[48]Barth, *Church Dogmatics*, vol. I, *The Doctrine of the Word of God*, trans. G. T. Thomson and H. Knight, ed. G. W. Bromiley and T. F. Torrance (Edinburgh: T & T Clark, 1936-1956), 1:363.

[49]See the astonishing *Church Dogmatics* II/1:§28.

[50]Ibid., III/1:69-71; IV/2:49-50, 91-115.

[51]Ibid., IV/1:§59.1.

[52]Ibid., IV/1:176-77; IV/2:42-44, 352-60.

[53]Ibid., IV/2:100.

God Is Agape

The Father, the Son and the Holy Spirit *are* only as they entail each other, only as they turn to each other. Thus, they are persons. The person of the Son goes out *from* and simultaneously goes out *to* the Father. As an other the Son turns to the Father in love and *is* loved. The love of God is outgoing. Creaturely being, through and through dependent upon God's care, is penetrated and embraced and affirmed as the Son moves outward to it, and thus destines it for God's kingdom.

Jesus of Nazareth is a passionate turning to this outgoing, to this eternal Word of address, a responsive "here I am." All that he is as a human being is given to the outgoing of God, to the coming of the kingdom. Jesus is expropriated but also glorified by it. He is simply given to the reign of God, to the love of God, to the coming of God. He is nothing but a face turned to God, a *vis-à-vis* that finally has only one visage. He is no substance, no self at the center of his world, no owner of his own properties; there is in him nothing standing beneath or behind. All that has been given up. Thus, before the face of God he is a transparent human face. Thus, he is a human person. Thus, he is hallowed. There is one face, one *prosōpon,* one *hypostasis,* one *persona* here: the concurrence of what human being is and wills with what God is and wills. But, of course, *this* person is nothing but a relation to the Father and the Spirit.

Union according to *ousia* occurs in the gifting of mutual dependence of the persons of the Trinity. Union according to *hypostasis* occurs in the enhypostatic hallowing of Jesus of Nazareth to the eternal personal address of God. We, however, are gifted, hallowed, by the embrace of this union, an embrace that is even now eschatological.[54] In Christ our destiny is given to be God's kingdom. God's "yes" to the lost, the poor, the suffering, the imprisoned, the condemned, the dying, the dead, the damned, is the eternal "yes" of God's coming rule. To be in Christ is to live from and toward what is to come, how-

---

[54]Panayiotis Nellas, *Deification in Christ: The Nature of the Human Person,* trans. Norman Russell (Crestwood, N.Y.: St. Vladimir's Seminary Press, 1987), p. 233.

ever fragmentarily. To be in Christ is to be a human person, however fragmentarily.

Life in Christ is shifted now to the life to come. We live in hope. The life to come is, however, no fixed, unbreathing state. It is a complex gathering of the shifting intersection points of relations of giving. Yet this gathering lacks an intrinsic center. It is what it is as it in turn moves from itself in adoration to the one enhypostatized in the outgoing face of God, a face that is no self-identical substance but God's eternal address, the second person of the Trinity, the one who *is* only as a relation of an other to the person of the Father and the person of the Holy Spirit. Thus, the kingdom adores the Trinity to whom it has been liberated.

It is therefore our destiny to look beyond ourselves, to lay down our defenses, to lose control, to face God and in so doing to face those whom God faces. It is our destiny to pray, to pray as an abandonment to God's love that is itself by looking to others *whom* God loves.

It is because God loves with the complexity of what the trinitarian persons entail that we, too, are persons, that we, too, go out to God and to one another. Therefore, to pray to God is never a lonely act, not even in this broken world. It is always the one who is destined to be eschatological love, the one whose self is to be her neighbor, the one who is as she points to each and to all—it is always this one who prays. To be a human person is to be by God's gift a passionate act of social prayer that says in one voice, "Here I am."

In the hearing of the gospel, in the regeneration of baptism, in the celebration of the Eucharist, in the act of mercy, in the quietness of meditation, in the ecstasy of joy, in the gift of justice, in the openness of faith, hope and love, the kingdom has begun to arrive. Although it is only in the end that we will have become persons in the truest sense, insofar as we live concretely here and now from the end, persons have begun to stir. It is thus not a complaint to say that "now we see in a mirror dimly," because our speech lives from the vital hope that "then we will see face to face."

3

Tacit Holiness

The Importance of Bodies & Habits in Doing Church

RODNEY CLAPP

Holiness is not a popular subject in contemporary theology. Among the vital and much investigated theological themes of the last few decades, one might list such topics as political liberation, religious epistemology, gender and gender language, and of course an overall preoccupation with methodology. But academic or "professional" theologians have devoted little attention to holiness. The same is true at the less rarefied level of seminary education. As John Alexander writes, seminary education, like secular education, has become a matter of information transfer. "If you told seminary professors that they were indistinguishable from professors in secular graduate schools, most would be delighted. They suppose God is a topic to be studied much as you would study geological formations on the moon. And from about the same distance." In fact,

> The holiness of professors . . . is of little concern. And it doesn't occur to anyone that the center of the curriculum should be teaching students to be holy. No, we settle for transferring information into their

> heads. And usually it isn't even information about holiness. Seminaries rarely offer courses with information on Mother Teresa, let alone courses on how to *become* Mother Teresa.[1]

Thus Stanley Hauerwas in his essay "The Sanctified Body" (pp. 19-38) proves himself distinctive. Here is a serious theologian willing to take up the subject of holiness. Even more, he is willing to suggest not merely that holiness be a subject of dispassionate study, but in fact that Christians should worry about becoming holy. In his essay Hauerwas makes it clear that he wants holiness, but holiness of a particular sort—a holiness that does not lose the "catholic character of the church" (p. 20). He doesn't say exactly what he means by *catholic*, but it is clear from his subsequent exposition that at least part of what he means is *catholic* in the sense of the church universal and whole, and accordingly a Christianity with a deep appreciation for the intrinsically corporate—both bodily and socially—nature of its holiness.

Of course, holiness has not been neglected in recent theology for entirely specious reasons. In the North American evangelical churches holiness has been grossly trivialized by its exclusive obsession with "personal" abstinence from habits and pastimes that not even many Christians would consider vices. Under this paradigm the avoidance of social dancing, movie-viewing, cardplaying, alcohol and tobacco becomes the primary mark of holiness. Though I suspect Hauerwas might be more patient with such behavioral codes than those who have had to live under them, the holiness he is after is decidedly not this individualized and privatized piety. Instead, Hauerwas most admires the sort of holiness exemplified in peasant Roman Catholicism. He means nothing pejorative by the appellation *peasant* but appreciates the peasant as someone who "works every day at those duties necessary for us to eat, to have shelter, to sustain the having of children and to carry on the basic practices necessary to sustain communities" (p. 21). Given their formation and material situ-

---

[1]John F. Alexander, *The Secular Squeeze* (Downers Grove, Ill.: InterVarsity Press, 1993), p. 31.

ation, "peasants may not be 'intellectuals', but they have knowledge habituated in their bodies that must be passed on from one generation to another" (p. 21).

Consequently, peasant Catholicism does not focus on being a Christian or becoming holy by mere mental assent to beliefs or doctrines. Beliefs and doctrines are important but peasants, not being intellectuals, are often not skilled at articulating or defending them—that they leave to bishops and theologians. The primary knowledge of peasant Catholics, then, is not analytical, detached, highly rational and systematized. It is instead the knowledge of embodiment and habituation, a knowledge intended "to have one's body shaped, one's habits determined, in such a manner that the worship of God is unavoidable" (p. 22).

Tacit Holiness

It helps me, at least, to better understand what is meant by embodied and habituated knowledge by referring to Michael Polanyi's concept of "tacit knowledge." Polanyi observed that a great deal of knowledge (even that of intellectuals) is not consciously known and used. An able physicist might, for example, analyze successful bicycle riding and the concomitant physics of balance. The physicist might then arrive at the formula that a person wobbling on a moving bicycle should curve in the direction of the imbalance to a degree inversely proportional to the square of the speed at which the bicycle is going.[2] The formula is correct in its own way. Indeed, people who right the balance of their bicycles do curve to a degree inversely proportional to the square of their speed. But does knowing the formula or being capable of its calculation make one a better bicycle rider? Surely not, as any number of six-year-olds might ride circles around the eminent physicist who has never before set feet to pedals.

And in fact no six-year-old I know of was ever taught to ride by first studying the physics of bicycling balance. Instead, parents put

---

[2]Michael Polanyi, *Personal Knowledge* (Chicago: University of Chicago Press, 1962), p. 50.

their children on bicycles, walk beside them, push, catch and exhort with simple advice such as "Don't turn too sharply," "Keep up some speed," and "Watch out for that tree!" Eventually, via coaching, intuition and actual trial and error, children consistently learn to ride bikes. What they have achieved is embodied and habituated knowledge—knowledge they likely cannot articulate or reflect on in analytical fashion, but very real and effective knowledge nonetheless. They are not thinking about the pedals or handlebars as they ride but are tacitly aware of them, just as the blind person is tacitly aware of his cane as he concentrates on what lies ahead or as the pianist is tacitly aware of the keyboard as she concentrates on sheet music.

And many types of knowledge that involve our bodies are likewise tacit knowledge. Swimmers typically do not know the pulmonary physiology that makes their bodies float. As Stanley Fish points out in a marvelous essay, baseball players can rarely as illuminatingly talk about playing baseball as can announcers: "playing baseball" and "explaining playing baseball" are distinct activities.[3] And a friend who was formerly a dancer tells me that her instructor frequently insisted, "The worst thing you can do while you're dancing is to think about it."

What Hauerwas wants to do, then, is to remind us that Christian knowledge—firsthand knowledge *of* holiness and not merely *about* holiness—involves the body and habituation. Like Polanyi, he challenges the assumption that we do not need our bodies for all kinds of knowing. Hauerwas's (and Polanyi's) move is postmodern in the regard that it rejects the (modernist) Cartesian assumption that to know something we must achieve "an absolutely clear and indubitable conceptual grasp of an object of knowledge."[4] In fact, not even analytical and scientific knowledge is known all at once, and comprehensively, or achieved apart from bodily senses. As theologian Colin Gunton glosses Polanyi, "Although knowledge involves concepts, no

[3]Stanley Fish, "Dennis Martinez and the Uses of Theory," in *Doing What Comes Naturally* (Durham, N.C.: Duke University Press, 1989), pp. 372-98.

[4]Colin E. Gunton, *Enlightenment and Alienation* (Grand Rapids, Mich.: Eerdmans, 1985), p. 38.

knowledge can be of such a kind that it is completely explicit or exhaustive." Consequently, we need to trade an "excessive regard for explicit [rationalized, formalized, abstracted and articulated] knowledge" for the recognition that "always 'we know more than we can tell.'"[5]

As regards holiness and so much else, peasants always know more than they can tell. They know holiness "in their gut" or "in their bones," bodily and tacitly, through the tutelage of habit. Such tacit holiness is reflected in classical writers on Christian spirituality like Thomas à Kempis, who said, "What good do you get by disputing learnedly about the Trinity, if you are lacking in humility and are therefore displeasing to the Trinity?" and, "I would rather feel compunction than know how to define it."[6]

The Illusions of Individualism

Of course, Hauerwas recognizes that many (if not most) Christians in modern, highly Cartesian, "developed countries" are far from embracing this sort of bodily, social, tacit holiness. Instead, as he puts it, we "believe our lives are the outcome of choices we have made" (p. 22). We are believers in individual autonomy, in the self-made man or woman who abstracts from his or her particular bodily, social and historical situation, and with careful reflection on any number of options, decides to be who he or she wants to be.

But it is important to recognize that this understanding of self as individuated, isolated and in control of its own destiny is, ironically, a social and historical creation. In the span of recorded history it is in fact a highly unusual and even unique interpretation of the self. As rhetorician Wayne Booth notes, the self as "in-dividual" (literally "undivided one") is barely more than two centuries old. The in-dividual was invented by a succession of Enlightenment thinkers and was reflected, in its most extreme but perhaps also its most widespread interpretations, in a view of the self as "a single atomic isolate,

---

[5]Ibid., pp. 38-39.

[6]Thomas à Kempis, quoted in Alexander, *Secular Squeeze*, p. 31.

bounded by the skin, its chief value residing precisely in some core of individuality, of difference."[7]

Thus it remains popular—almost second nature—to think we get at our "true self" by peeling away social ties like the skin of an onion. The "real me" is not my membership in the worldwide church, my shared kin with Clapps around the country, not my connection—with three million other people—to the geography and culture of Chicago. The "real me" is my unique, in-dividual, core self. The in-dividual self values itself most for what is supposedly utterly different and unconnected about it. But, objects Booth, such an understanding of self is incoherent. Can we really believe that we are not, to the core, who we are because of our kin, our occupations, our political and social situations, our faith or philosophical communities, our friendships? And if our "true self" is whatever stands apart from those around us and is altogether unique about us, most of us are in trouble. The bizarre modern notion of the self means even the greatest geniuses have only minimal worth. "Goethe," says Booth, "was fond of saying that only about 2 percent of his thought was original."[8] Truly, as Philip Slater remarks, "The notion that people begin as separate individuals, who then march out and connect themselves with others, is one of the most dazzling bits of self-mystification in the history of the species."[9]

In fact, Booth continues, "People in all previous cultures were not seen as essentially independent, isolated units with totally independent values; rather, they were mysteriously complex persons overlapping with other persons in ways that made it legitimate to enforce certain kinds of responsibility to the community." In these settings persons were not " 'individuals' at all but overlapping members one of another. Anyone in those cultures thinking words like 'I' and 'mine' thought them as inescapably loaded with plurality: 'I' could not even

---

[7]Wayne Booth, "Individualism and the Mystery of the Social Self," in *Freedom and Interpretation*, ed. Barbara Johnson (New York: BasicBooks, 1993), p. 81.

[8]Ibid., pp. 87-88.

[9]Philip Slater, *The Earth Walk*, quoted without further attribution in Lawrence Stone, *The Past and the Present Revisited* (London: Routledge and Kegan Paul, 1987), p. 325.

think of 'my' self as separated from my multiple affiliations: my family, my tribe, my city-state, my feudal domain, my people."[10]

In light of Hauerwas's concern for a bodily and social holiness, as opposed to an individuated holiness, it is worth asking: Are the biblical cultures part of the "previous cultures" Booth here remarks on? Scholars have again and again noted the Hebrew conception of "corporate personality," the understanding that families, cities, tribes and nations possess distinctive personalities and that individuals derive identity from and so might represent these social bodies.[11] We need no new frame when we extend this picture into the Greco-Roman world of New Testament times. Writing on the concept of personhood in the New Testament setting, Bruce J. Malina notes, "the first-century Mediterranean person did not share or comprehend our idea of an 'individual' at all." Rather, "our first-century person would perceive himself as a distinctive whole set in relation to other such wholes and set within a given social and natural background."[12]

Thus when Paul spoke of the church as a "body," he borrowed the metaphor from a fable widely used in several cultures of antiquity. Just as "Israel" could serve as the name either of an individual (Jacob) or a community (the nation), so could Paul use "Christ" to refer to an individual (Jesus the Messiah) or a community (the church). In the words of New Testament scholar Charles Talbert, "'Members' . . . is Paul's term for the parts of the body through which the life of the body is expressed (see 1 Cor 12:12, 14-26; Rom 6:13). Paul is saying then that individual Christians in their corporeal existence are the various body parts of the corporate personality of Christ through which the life of Christ is expressed."[13]

---

[10]Booth, "Individualism," pp. 78, 79.

[11]For a classic statement, see H. Wheeler Robinson, "Hebrew Psychology," in *The People and the Book*, ed. A. S. Peake (London: Oxford University Press, 1925), pp. 353-82. Robinson remarks that such doctrines as original sin are incomprehensible without a notion such as corporate personality. I would add that our thoroughgoing individualism also threatens to render incoherent the doctrines of atonement, of the church and even, most fundamentally, the Trinity.

[12]Bruce J. Malina, *The New Testament World: Insights from Cultural Anthropology* (Atlanta: John Knox Press, 1981), pp. 54, 55 (emphasis in original).

[13]Charles Talbert, *Reading Corinthians: A Literary and Theological Commentary on*

All this confirms Hauerwas's adaptation of Dale Martin's work on *The Corinthian Body*. At the beginning of his book, Martin in fact sets his interpretation of the Biblical self in direct opposition to the Cartesian, individuated self. Martin writes that Paul (and the ancient world in general) simply knew nothing of such dichotomies as matter versus nonmatter, corporeal versus psychological, spiritual versus physical, and so on. It was Descartes who practically "invented the category of 'nature' as a closed, self-contained system, over against which he could oppose mind, soul, the spiritual, the psychological, and the divine." Descartes removed to the nonphysical realm all those aspects of reality that exercise volition and true freedom—that is, God, and the soul or mind or "I" of the human self. He wrote that "that 'I,' that is to say, the mind by which I am what I am, is wholly distinct from the body" and thus categorized the human body with other nonvolitional entities, namely those comprising what he called nature. "It was this disenchanted, machine-like nature that could be legitimately examined and manipulated scientifically."[14]

Thus, taking into account Descartes and a long line of Enlightenment thinkers who followed him, it seems clear that modern individualism emphasizes the noncontingency and hence the self-determination of the nonphysical, essential self. The body, as well as all others things nonvolitional and merely physical, is nonessential and instrumental, to be directed and used by the deeper, real self. Control, then, is a much-prized modern value. Rationalism bolsters this sense

---

1 and 2 Corinthians (New York: Crossroad, 1987), p. 31. Also J. Paul Sampley: "Paul thinks of believers' relationship with Christ in terms of solidarity with, participation in, or belonging to Christ. . . . Those who have faith are one together in Christ. This solidarity with Christ is Paul's primary identification of believers."

He adds, "Just as surely as one does not snub the workings of the Spirit, one does not disregard the community in one's life of faith."

And: "Paul's great interest in the health and growth of the individual's faith is always set within his concern for the well-being of the community, and his commitment to community is always located within his conviction that God's renewal of the entire cosmos is under way" (J. Paul Sampley, *Walking Between the Times: Paul's Moral Reasoning* [Minneapolis: Fortress, 1991], pp. 12, 43, 118).

[14]Dale Martin, *The Corinthian Body* (New Haven, Conn.: Yale University Press, 1985), pp. 4-5.

of control by removing us from the limits and vulnerabilities of the body, something we clearly did not autonomously choose, and which possesses a nastily persistent vulnerability to sicknesses, injuries and deaths we cannot predetermine.

It is exactly this sense of control and self-determination that Hauerwas fears sets Christians in "advanced" countries off from peasants. Contemporary professionals especially, heir to the "endemic individualism and rationalism of modernity" (p. 23), believe themselves to be their own masters. But this sense of control and self-determination is, as I have indicated, historically anomalous. Even non-peasants, in all times prior to ours, did not imagine that they controlled their destiny. Professionals, and the modern middle class they so much comprise and influence, are enrapt by an illusion of individualistic self-determination that ironically makes them less free than peasants. As Hauerwas puts it, "Peasants know there are masters and accordingly develop modes of resistance. Professionals think they are masters, which of course makes them completely incapable of defending themselves" from the actual masters of their destiny (p. 22 n. 5). You cannot effectively resist masters you do not recognize as such, and we modern individualists are constantly taught and coached to buy into the illusion of self-determination, but not to notice that we are so taught and coached by a pervasive mass, technological, liberal, consumer capitalistic system.[15] Like Hauerwas, the Catholic writer Robert Inchausti worries about a modern, professionalized middle class whose very "longing for a more profound existence makes them prey to salesmen and commercial visionaries who sell them dreams without inspiring them to the discipline necessary for authentic transcendence. They become consumers—systematically miseducated, underestimated, financially pampered, and morally exploited. They are nonpersons who are celebrated by the media as existentially free but whose history, ethnicity, personality, and moral seriousness have been marginalized, if not entirely debunked, by a

---

[15]For unpacking of what I mean by such a designation, see the remarks on mass-techno-liberal-capitalism in chapter 12 of Rodney Clapp, *A Peculiar People: The Church as Culture in a Post-Christian Society* (Downers Grove, Ill.: InterVarsity Press, 1996).

world socio-economic order that runs according to its own impersonal rules and agenda."[16]

In such a setting Hauerwas is right to think that the church faces a considerable challenge in becoming a "disciplined body of disciples" (p. 23). He is also correct to recognize the professionalized middle class as a special challenge to a disciplined church, since professionals and consumers "assume the church has no right to tell them what to believe or do. They assume what they know as a banker or lawyer or university professor gives them critical perspective on Christianity. Too often they get to make up the kind of Christianity they like" (p. 22 n 5). How, then, to recover a disciplined, embodied and habituated church in a setting so much more congenial to an individualized, rationalized, self-determining consumer "Christianity"?

Suffering and the Embodiment of Holiness

I have, of course, already begun to answer this question. To recognize that the isolated, self-creating self is a creation of modernity, and in fact historically anomalous, is to see that this rendering of self is at least open to challenge. That challenge can in turn lead to closer reflection on whether or not it is at all coherent to believe in—let alone try to be—a "self-creating self." Further, I have mentioned Hauerwas's appropriation of the work of Dale Martin, which suggests that biblical anthropology countenanced no such conception of the self. For Christians, this is a serious recognition indeed. What if Martin is right that for the apostle Paul "individual bodies have reality only insofar as they are identified with some greater cosmic reality," so that "Christian [physical] bodies have no integral individuality about them" but in truth have their "identity established by participation in a larger

[16]Robert Inchausti, *The Ignorant Perfection of Ordinary People* (Albany, N.Y.: State University Press of New York, 1991), p. 141. Inchausti's excellent book examines such "religious plebeian" leaders as Gandhi, Solzhenitsyn, Elie Wiesel, Mother Teresa, Martin Luther King and Lech Walesa who have, out of their religious traditions, recognized the reality of the inimical mass modern system, and called for antidotes that include an "elevated sense of the self as socially and historically constituted and therefore inherently politically engaged and morally responsible" (p. 141).

entity" that is the "body of Christ"?[17] Then Christians may have little investment in trying to construct for themselves modern, Cartesian selves. We may rather turn to the difficult hermeneutical and practical tasks of becoming Christian selves via our participation in the church as the disciplined, habituated and habituating body of Christ.

For an example of this task, Hauerwas resorts to Arthur Frank's account of illness and suffering in *The Wounded Storyteller*. Following Frank, Hauerwas notes, "Sickness makes it impossible to avoid the reality of our bodies" (p. 29). However much we as good moderns might like, generally, to ignore our contingency and vulnerability by Cartesian abstraction, "When I am sick I am not a mind with a suffering body, but I am the suffering body. Illness may then be the only time that we have the opportunity to discover that we are part of a story that we did not make up" (p. 29).

Of course, this is the case only when we are seriously ill and cannot escape or eliminate our suffering by the ministrations of modern medicine. When my headache can be extinguished by a prescription or my jutting jaw resculpted by plastic surgery, I am only encouraged in the modern myth that Cartesian science can enable me to escape or eliminate affliction and remake my self at will. Accordingly, as medical ethicists have often observed, modern medicine has concentrated on curing patients to the serious neglect of caring for patients. This medicine, in Hauerwas's account, demands a body that "is characterized by self-control, through which it seeks predictability by employing therapeutic regimens. Such regimens organize the body in the hopes that contingency can be avoided or at least compensated" (pp. 30-31). Modern medicine thus helps us deny contingency by striving to eliminate its undeniable manifestation in acute suffering, that is, by offering cures (the more immediate the better), and by largely removing incurable suffering to (and hiding it in) institutional settings. Hence the likelihood that most of us will die in hospitals, not in our homes.

The state of such medicine is all too accurately satirized in Sidney Lumet's 1997 film *Critical Care*. Lumet depicts a young Dr. Ernst, just

---

[17]Martin, *Corinthian Body*, pp. 131-32.

finishing his residency. Ernst longs to study under a senior doctor who uses computer monitoring and advanced technology to "treat" his patients from a centralized laboratory, dispatching nurses to administer drugs and other therapies, and dreaming of the day when he will never have to actually see or touch a patient's body again. As the story progresses it becomes increasingly clear to Dr. Ernst that he is caught up in a corrupt practice of medicine in which doctors, lawyers and insurance companies vie to garner more money and prestige. All this is done at the expense of patients who are no longer identifiable persons but merely cited by their bed numbers—hospitals make money, after all, by filling their beds. Only when he is forced through circumstances does Dr. Ernst realize that he has stopped paying any mind to the patients, to the people who are suffering under and to some extent because of his care. At the film's end he is leaving the hospital when a rollerblading teenager crashes in front of him. As he rushes to the boy's aid, a mentor doctor shouts across the parking lot that he'd better be careful. He might get sued—and he doesn't even know if the boy has insurance. Ernst begins to tend to the boy anyway. When the boy asks, "Are you a doctor?" Ernst, with a pause, realizes that finally he can say with real truth, "Yes, I am a doctor."

What the film, coupled with Hauerwas's and Frank's reflection, suggests is that even the most idealistic, well-intentioned physician can be only as good as the practice in which he is bodily and socially embedded allows. If serious illness can still force us to admit our vulnerability and contingency, Christians should be among those who imagine that medicine can still be about caring and bearing suffering, and not merely, or even especially, about curing. As persons who believe they are creatures of the God of Israel, and not of their own creation, Christians need not pretend to control their own lives and deaths. As Hauerwas writes, "This requires that we understand that heroism is not to be identified with those who can 'do something,' but rather is to be found in those who persevere through suffering. What such a body offers is not victory, but testimony. Our healing is not the overcoming of our illnesses, but rather our ability to share our going on with one another through the community stories create" (p. 33).

Consequently the church should be concerned with a renarration and reembodiment of illness and the objectives of a genuinely Christian medicine. Pastors and Christian patients should not, for instance, regard bedside prayers as desperate pleas to be thrown up once doctors admit there is "nothing more to be done." Instead, such prayers are part of enabling the patient to own her illness and her fragile body, to have herself and her loved ones depend on the care of the church, and to witness to a hope that does not finally despair even in the face of death. Occasions of such suffering are also times of profound tacit knowledge, of knowing more than we can ever tell. Who, after all, standing at the bedside of a gravely ill person, has not been at a loss for any adequate words? Surely the sick body "eludes language" (p. 29). Yet at the same time bodily presence at the bedside—and the patient's allowance of it—communicates care and solidarity. Very often our deepest and best communication is the holding of a hand or the rubbing of a shoulder or the drying of a tear. In such acts we embody the way and the spirit of holiness.

The (Double) Body at Worship

Illness shatters the illusions of independence and noncontingency constantly promoted by mass, technological, liberal and consumer capitalist society. Yet in looking to serious illness, Hauerwas is concerned to recover truths about our whole lives, not merely about who or what we are when we are sick. Hauerwas wants Christians to notice this, intellectually and in such practices as health care, and to carry these lessons into other realms of our existence. Hard as it definitely is to bear suffering or to stand by the suffering of others, it may be even more of a challenge to take an awakened awareness of creaturely contingency out of the hospital and back into the rest of our lives. The truth, in the Christian confession, is that we are always contingent and dependent, whether in sickness or in health, in poverty or in wealth, in death or in life.

I agree with Hauerwas and now want to emphasize a reality that is more regular and formative of Christian character. Even when Christians are not sick, we are routinely called to worship. Nowhere, I think, should we be so profoundly habituated in our creaturely dependence on God and God's grace as we are at worship.

Worship, most fundamentally, is not something we initiate or create. We are gathered to common worship at the initiative of the Father, through the saving deeds of the Son, in the power of the Holy Spirit. In common or corporate worship we celebrate and rehearse, at least weekly, the work of Yahweh in the story of Israel and Jesus Christ. Had not God first acted in Israel and Jesus, we would have no creation and redemption to celebrate, no Christian story to rehearse. And if God the Spirit did not call Christians to gather, we would not leave our beds each Sunday morning—each "little Easter"—to be and to act as members of Christ's body and of one another.

In worship, then, we act out of contingency. We recall that our very being is contingent; this is exactly what it means to be "creatures" as opposed to self-made men and women. We bless God for our creation, our redemption and our sustenance in the faith. And our whole selves—body, mind and soul—are required for genuine worship. We gather bodily to pray, to sing praises, to lay hands on the sick, to baptize and to celebrate the Eucharist. And all of this is initiated and enabled by God. So we profess our contingency in word and also in our very physical presence, one to another, at worship. No single one of us, no "in-dividual," constitutes Christ's body. We do so only by gathering and sharing and depending on the gifts God has given the community called church. Not all are teachers, not all counselors, not all prophets or priests. We need one another to be what we are, the body of Christ. God's Spirit gifts each of us not for any individual's sake, but "for the common good" of that body (1 Cor 12:7).

So Christian worship is the preeminent practice of the kind of holiness Hauerwas seeks—a doubly embodied holiness, corporate both physically and socially. Consider the sacrament of baptism. Baptism immerses our physical bodies in the body of Christ. In so doing, it initiates us into a social body in which "there is now neither Jew nor Greek, slave nor free, male nor female, for all are one in Christ Jesus

(Gal 3:28) and it grants us Gentiles a new citizenship in the commonwealth of Israel (Eph 2).

In baptism as in other sacraments, physical and social bodily involvement is crucial. The baptizand's physical body, not just his or her mind or soul (whatever that might be), must be touched and wetted. And the baptizand must be surrounded and joined by the social body of the church—private baptisms are an oxymoron imaginable only in the distorting wake of modern individualism. The social body vows (as in my own Episcopal tradition) to "do all in your powers to support this person in his or her life in Christ." In these and other ways it necessarily participates in the individual's baptism. Certainly baptism involves words and rational assent. But the words of the baptismal formula and our rational assent are never separated from bodily and tacit knowledge. In fact, the tacit knowledge of baptism both enriches and in some ways goes beyond intellectual knowledge. Baptismal water, for instance, teems with tacit significance. It palpably hints at creation (water is life-giving), creaturely mortality (water drowns and disintegrates), cleansings, new life and much else. So, as Polanyi would have it, the one who has been baptized can never tell all he or she knows through baptism, through the actual participation of the physical and social body in this rite. Accordingly, one can no more truly "know" baptism without bodily participating in a baptism than one can really know the ecstasies of sex by simply reading a therapist's manual. It is never simply or even primarily a matter of intellectual or soulful consent. As the church father Tertullian remarked,

> [In baptism] the flesh is washed in order that the soul may be cleansed; the flesh is anointed in order that the soul may be consecrated; the flesh is signed in order that the soul may be fortified; the flesh is overshadowed by the imposition of the hand in order that the soul may be illumined by the Spirit. . . . Therefore, those things which work together are not able to be separated in reward.[18]

[18]Tertullian *De Resurrectione Carnis* 8.6-12, as cited in Mary Timothy Prokes, FSE, *Toward a Theology of the Body* (Grand Rapids, Mich.: Eerdmans, 1996), p. 137.

Similarly, the Eucharist for its enactment requires double embodiment, both physical and social corporateness. For it, too, the body of Christ must gather and the physical bodies of the members of Christ must go through certain motions and assume particular positions. Kneeling and standing at the appropriate junctions, cupping one's hands to receive the bread, crossing one's self—all of these are bodily gestures communicating and reinforcing a wealth of tacit knowledge. To kneel, for example, is to admit our submission more powerfully and completely than any words alone might; to stretch out empty and receptive hands for the bread of life is to practice our contingency fully and dramatically. So the Eucharist leads us to say that corporality is the very mediation where faith takes on flesh and makes real the truth that inhabits it. "It says this to us with all the pragmatic force of a ritual expression that speaks by its actions and works through the word, the word-as-body. It tells us that the body, which is the whole word of humankind, is the unavoidable mediation where the Word of a God involved in the most human dimension of our humanity demands to be inscribed in order to make itself understood. Thus, it tells us that faith requires a *consent to the body.*"[19]

With the faithful consent to the body that is worship, we can begin to have healed the Cartesian separation of matter and nonmatter, of the corporeal and the psychological, the physical and the spiritual.

Unfortunately, much Protestant worship only serves to reinforce such Cartesian dichotomies. If the Eucharist, for instance, is mainly a Zwinglian matter of the individual's remembering and thinking about Christ's sacrifice, then the gathering with other Christians is secondary, dependence on God's initiatory grace is downplayed, and a rationalistic, nonbodily human action is made central. A merely memorialist Eucharist is double disembodiment, separating the worshiper both from the social body of the gathered church and from his or her own physical body. Such worship may, in fact, be more a Cartesian sort of disembodied *observation* than it is actual *participation*

[19]Louis-Marie Chauvet, *Symbol and Sacrament,* trans. Patrick Madigan, S.J., and Madeleine Beaumont (Collegeville, Minn.: Liturgical Press, 1995), p. 376.

in the body of Christ. This is yet another cause to join Hauerwas in his concern for a genuinely catholic Christianity. A full and real holiness is tacit holiness, necessarily habituated and (doubly) embodied. If some forms of Protestant worship fail to be less than catholic in this regard, then they are actually misshaping the Christian formation of their practitioners. And that is a serious matter indeed.

4

Holiness as the Renewal of the Image of God in the Individual & Society

THEODORE RUNYON

One of John Wesley's preferred ways of describing salvation was as "the renewal of the image of God." Again and again he returns to this theme, which the twentieth century's foremost Wesley scholar, Albert Outler, calls "the axial theme of Wesley's soteriology."[1] If it plays a key role in his doctrine of salvation, it follows that Wesley's understanding of the image of God must be important for his doctrine of holiness. But what does Wesley really mean by this term, *the image of God?*

Wesley's first "university sermon," preached at St. Mary's Church, Oxford, in 1730, when he was a tutor at Lincoln College, was titled "The Image of God" and gives us a glimpse into the meaning of the phrase for him. In this sermon he makes use of a metaphor stem-

[1]Frank Baker, ed., *The Works of John Wesley,* bicentennial ed. (Nashville: Abingdon, 1984), 2:185 n. 70. Cf. Theodore Runyon, *The New Creation: John Wesley's Theology Today* (Nashville: Abingdon, 1998), pp. 222-33.

ming originally from the Eastern fathers: humanity as the image of God is a *mirror that reflects* what it receives from God, reflects it both back to God and into the world. The image is not, therefore, something that humanity has or that is lodged within the human being but is an ongoing relation in which humanity receives and gives. And what is received is *love*. "The flame of [love] was continually streaming forth" directly to humanity from the divine Source and then by reflection back to God and outward to God's offspring who likewise bear "the image of their Creator."[2] This metaphor of humanity as a mirror is significant because it avoids the Western tendency to subjectivize the image by identifying it with some capacity the human being possesses as did Thomas Aquinas, who identified the image with reason, or Immanuel Kant, identified it with conscience. As a mirror the image is *relational*. It only functions as it is receiving and reflecting. Therefore, it is not a capacity within the creature because a mirror can only reflect something beyond itself. If in salvation the image is being renewed, the essential qualities of the image are to be found not within humanity but in that which humanity is called to reflect. It reflects love back to God, the love called forth by receiving God's love, and it reflects God's grace, justice, mercy and love to the neighbors God has given us. This is how, according to the favorite passage of the Eastern fathers that Wesley read to begin his day on May 24, 1738, Christians are to be made "partakers of the divine nature" (2 Pet 1:4). That is, the bearers of the image of God are made agents of God's redeeming power, that power by means of which, according to his "great and precious promises," he will lift up the fallen creation.

Wesley spells out the way in which this is accomplished in his Fourth Discourse on the Sermon on the Mount, which was a commentary on Matthew 5:13-16 but was at the same time an apology for the fundamentally *social* nature of Christianity against the quietistic and individualistic interpretations of some of the Moravians and of his former mentor, William Law, who advised those seeking to be true

---

[2]Baker, *Works of John Wesley* 4:295.

Christians to "'cease from all outward actions' [that is, from any good works]; wholly to withdraw from the world; to leave the body behind us; to abstract ourselves from all sensible things, . . . and, instead of busying ourselves at all about externals, we should only commune with God in our hearts." This, it was claimed, "is the far more excellent way, more perfective of the soul, as well as more acceptable to God."[3]

Wesley begins his response by laying down the thesis that "Christianity is essentially a social religion, and that to turn it into a solitary religion is indeed to destroy it. . . . When I say [Christianity] is essentially a social religion, I mean not only that it cannot subsist so well, but that it cannot subsist at all without society, without living and conversing with [others]."[4] The reason for this judgment is twofold: first, Christian faith comes to birth in a social context through contacts with other Christians, and second, Christian faith requires a social context to accomplish what God wants to accomplish through it. Wesley prefaces these remarks with a reference to the renewal of the image of God, and so it is clear that he presupposes the social character of the image.

As to the first reason for the judgment that Christianity is a social religion: faith emerges in a social context. Wesley does not want to claim that it is impossible for God to act upon the heart of a person in isolation, but his observation is that such action usually occurs by means of other persons. "Though it is God only [who] changes hearts, yet he generally doth it by man."[5] Faith is not only born in this kind of social encounter, it grows and is nourished in community. This was the genius of the Wesleyan societies, classes and bands. They provided the necessary communal setting within which faith could not only be generated but enriched and sustained. This is why Wesley was convinced that "conference" (i.e., conferring together) was for Christians a means of grace as together they sought the meaning of Scripture, prayed and shared the life experiences that both chal-

---

[3]Ibid., 1:532.
[4]Ibid., 1:533-34.
[5]Ibid., 1:546.

lenged and confirmed their faith. Wesley could not imagine Christians not benefiting from this kind of exchange and mutual support. And he observed that where this kind of support was not available, although the seed sowed in preaching might quickly spring up, it withered away just as quickly. From 1746 to 1748 Wesley experimented with placing the emphasis on preaching alone without forming societies—with disastrous results. "Almost all the seed has fallen by the wayside; there is scarce any fruit remaining," he noted in the *Minutes* of the Conference of 1748. And at that conference the decision was made to turn again to the formation of societies.[6] Wesley's Calvinist friend and colleague, George Whitefield, consistently drew with his preaching larger crowds and more public acclaim than did Wesley. But Whitefield, after he parted ways with Wesley, did not give great attention to the infrastructure needed to continue the nurture of those who were attracted by his preaching and, as a result, the fruits of his labors were not as well-preserved.

Statistical research by Thomas Albin on the spiritual lives of 535 early British Methodists whose spiritual biographies were published in the pages of the *Arminian Magazine* and the *Methodist Magazine* shows that, according to their own testimony, only one-fourth experienced new birth in the context of preaching they heard prior to joining a Methodist society. By far the majority needed the nurture of the society, classes and bands, and spent an average of 2.3 years in this nurturing process before experiencing what they identified as new birth. In this process fellow class members, class leaders and lay preachers were the primary influences.[7] Thus, it was his recognition of the social character of Christian faith that required Wesley to invest the time and energy to build the organizational structures necessary to conserve and multiply the impact of the revival by ensuring social interchange, for Christian faith comes to birth and is nurtured through other persons.

---

[6]Richard P. Heitzenrater, *Wesley and the People Called Methodist* (Nashville: Abingdon, 1995), pp. 163-65.

[7]Thomas Albin, "An Empirical Study of Early Methodist Spirituality," in *Wesleyan Theology Today*, ed. Theodore Runyon (Nashville: Kingswood Books, 1985), p. 278.

As to the second and equally important reason for the judgment that Christianity is a social religion: the fact that the purposes of God are only fulfilled as those who are reconciled with God serve as "channels of grace" to other persons. The social nature of holiness is not only written into the origins of the faith relationship but also into the ongoing direction of it as well. The function of an image of God is to reflect and pass on the grace that is received to a world that can be healed only by encountering this grace. To be sure, God could renew the world by fiat. The Almighty could simply "act irresistibly, and the thing is done; yea, with just the same ease as when 'God said, Let there be light; and there was light.'"[8] But this is not the course God has chosen. Instead, God seeks to spread his reign through renewing his image in the hearts of human beings. And this requires a social process that without coercion honors the freedom of humans to respond to the promptings of the Spirit. God patiently allows "the little leaven [to] spread wider and wider."[9] "The kingdom of God will not 'come with observation,' but will silently increase wherever it is set up, and spread from heart to heart, from house to house, from town to town, from one kingdom to another."[10] But this is a process that must begin with the renewal of the image in those who are to be God's *agents* in the world. "This is the great reason why the providence of God has so mingled you together with other men, that whatever grace you have received of God may through you be communicated to others."[11] The Christian is enlisted as an agent in God's great undertaking of redeeming the nations. This is why "to turn Christianity into a solitary religion is to destroy it," for it can no longer function as it was designed to function, to communicate renewing grace to others.

Wesley provides a convincing illustration of this point. A necessary branch of true Christianity is peacemaking. As Matthew 5:9 testifies, this calling is "inserted into the original plan [Christ] has laid down of

---

[8]Baker, *Works of John Wesley,* 2:488.
[9]Ibid., 2:491.
[10]Ibid., 2:493.
[11]Ibid., 1:537.

the fundamentals of his religion." But peacemaking is not possible unless one is willing to enter into situations of conflict.[12] Hence, reasons Wesley, there is "no advice [in Scripture] to separate wholly, even from wicked men." Count not the wicked man as an enemy, says the apostle Paul, "'but admonish him as a brother,' plainly showing that even in such a case we are not to renounce all fellowship."[13] For it is precisely through communication that the image of God is able to mediate the power of God that has been received, "that every holy temper, and word, and work of yours, may have an influence on [others] also. By this means a check will in some measure be given to the corruption which is in the world."[14]

The analogies that Jesus employs in the Sermon on the Mount reinforce this same point of the social end toward which Christian faith is directed. "'Ye are the salt of the earth.' It is your very nature," says Wesley, "to season whatever is round about you. It is the nature of the divine savor which is in you to spread to whatsoever you touch; to diffuse itself on every side, to all those among whom you are."[15] "Ye are the light of the world. A city set upon an hill cannot be hid." Nor should it be. It has received light in order to share light. "Never therefore let it enter into the heart of him whom God hath renewed in the spirit of his mind to hide that light, to keep his religion to himself, especially considering it is not only impossible to conceal true Christianity, but likewise absolutely contrary to the design of the great Author of it." "Your holiness makes you as conspicuous as the sun in the midst of heaven." For what is received that makes holiness possible is love, and "love cannot be hid any more than light; . . . least of all when it shines forth in action."[16]

The social nature of holiness becomes even more evident when we turn to a later sermon of Wesley, "The New Birth," in which he describes in more detail the nature of the image of God. In this ser-

---

[12]Ibid., 1:534.
[13]Ibid., 1:536.
[14]Ibid., 1:537.
[15]Ibid.
[16]Ibid., 1:539-40.

mon he amplifies his notion of the image to include three factors: the *natural* image, the *political* image and the *moral* image. The natural comes closest to describing the image in the traditional Western way, namely those endowments with which the creature is blessed that make us "capable of God," spirits able to enter into conscious relationship with God. As spirit the natural image is endued with *reason, will and freedom*. But Wesley is more modest in what he ascribes to reason than were most Enlightenment thinkers. The metaphysical and mythological qualities of reason are replaced by functional ones. Yet these functions are of utmost importance, for reason operates in three ways necessary for any reflection: in *perception*, "conceiving a thing in the mind"; in *judgment*, comparing perceptions with each other; and in *discourse*, the "progress of the mind from one judgment to another."[17] Thus reason enables us to grasp how things work together, which makes it possible to discern order and relations, and to make right judgments. Moreover, reason can be of inestimable value for faith. Without reason it would be impossible to explain the basic principles of faith as found in the Scriptures and expressed in the creeds. It is reason that enables us to understand the nature of our relation to God and the way of salvation, as well as the implications of faith for life. But this proper use of reason is dependent upon the proper relationship of the image to God. Where the image is disfigured or the relationship undermined, however, human reason is likewise distorted so that it serves the creature and not the Creator. We ingeniously put our reason to work to rationalize and excuse disobedience and sin.

Two further marks that serve as characteristics of the natural image are will and freedom. These go together because Wesley recognizes that the human will has been corrupted by the Fall. Human disobedience has disrupted the relationship between the image and God so that the natural tendency of the human will is to be self-seeking. The fallen will is in bondage to sin. Yet if this bondage is complete and the will lacks any freedom, it cannot be held morally accountable. If

---

[17]Ibid., 2:590.

God's judgment is to be just, therefore, a degree of free will is necessary. "A mere machine is not capable of being either acquitted or condemned. Justice cannot punish a stone for falling to the ground." Without freedom "both the will and the understanding would have been entirely useless" because the capacity for agency—the ability to initiate and pursue objectives—would not be present. Therefore, Wesley is convinced that a loving God intervenes to introduce through prevenient grace "a measure of freedom in every man," which also gives rise to the universal phenomenon of conscience. Unlike Kant later, however, Wesley is at pains to say that this free will is not natural. "Natural free-will, in the present state of mankind, I do not understand."[18] But he does understand that a measure of free will is given as a divine gift to restore the fallen creature to responsibility and agency, and to open up the possibility of the renewal of the divine image. The gift of freedom gives to the conscience sensitivity and to the will the power to choose the good and resist evil.

Thus reason, will and freedom are primary characteristics of that spiritual being who bears the natural image of God. And in his description of the natural image Wesley comes closest to the traditional notion of the image as capacities that humanity possesses. Yet even here it is evident that these are not innate capacities so much as gifts given to enable human beings to carry out their calling to image and reflect their Creator, gifts that flourish when used in ways consistent with the will of the Giver, but gifts that also are easily distorted when turned to serve the selfish interests of the creature. These capacities are not neutral, therefore, but derive their character from the quality of the relationships in which they are employed. Holiness would be marked by the full and rigorous use of these gifts within a faithful relationship to the Giver.

The political image is the second basic way in which humanity reflects its Maker. God endowed this creature with faculties for leadership and management, to be "vicegerent upon earth, the prince and governor of this lower world." To humanity was given the special

---

[18]Ibid., 2:424-26.

responsibility of being "the channel of conveyance" between the Creator and the rest of creation, so that "all the blessings of God flowed through him" to the other creatures.[19] Thus humanity is the image of God insofar as the benevolence of God is reflected in human actions toward the rest of creation. This role as steward and caretaker of creation presupposes a continuing faithfulness to the order of the Creator. On this basis alone can humanity expect to maintain order in the world under its management. In our own time the implications of humanity's role as the political image have become ever more crucial not only with regard to other creatures and their survival but with regard to the whole environment. Yet here again, the quality of the image in its political calling is dependent upon the quality of the relationship of the "prince and governor" to the Creator. And holiness cannot be defined apart from the faithfulness of the political image, as steward and channel of the conveyance of blessings, to the Lord of all creation.

The third characteristic of the image of God is the *moral image.* This is the prime mark of the human relationship to God, according to Wesley, but also the one most easily distorted. The *natural* image consists of endowments most of which are retained in humanity, albeit in adulterated form, after the Fall. The *political* image is one that humanity continues to exercise, albeit in corrupted fashion, reflecting not the Creator but the pride, selfishness and insecurity of the human condition in a fallen world. But the *moral* image is neither a capacity within humanity nor a function that can be employed independently of the Creator, because it consists precisely in a relationship in which the creature receives continuously from the Creator and mediates further what is received. The lifeline for this is what Wesley calls "spiritual respiration":

> God's breathing into the soul, and the soul's breathing back what it first receives from God; a continual action of God upon the soul, the re-action of the soul upon God; an unceasing presence of God, the loving pardoning God, manifested to the heart, and perceived by faith;

[19]Ibid., 2:440.

> and an unceasing return of love, praise, and prayer, offering up all the thoughts of our hearts, all the words of our tongues, all the works of our hands, all our body, soul, and spirit, to be an holy sacrifice, acceptable unto God in Christ Jesus.[20]

This is the classic description of holiness, and it is foundational to any Wesleyan doctrine of holiness. For the moral image is the relational base on which the natural and the political image are set. Yet we need to remind ourselves that the moral image is only one part of Wesley's tripartite description of the image of God. And it is the renewal of the image of God in all its dimensions that is the goal of salvation and the fulfillment of the goal of Christian perfection. The moral image is basic, and it is the presupposition of the proper practice of the other two. But holiness has been defined too narrowly if we have left out of the configuration the natural and the political image. If we include them we include holiness not only of heart but of will and intellect as well, holiness not only of mind but of political and social responsibility and stewardship as well. In this way holiness takes on the social and cosmic dimensions that it had for the Eastern fathers who influenced and shaped Wesley's own position, dimensions too often missing from holiness thinking in the past.

To this we are called, as those being renewed in the image of God, to take into ourselves continuously that breath of life that comes from the Spirit of God and continuously to breathe out this same breath in a life of service to God, our fellow human beings, and to all creation. To humanity is given this crucial role as the natural image, the political image, and the moral image, mirroring and reflecting the Creator and mediating divine blessings to the world of nature and humanity around us. And this is holiness!

---

[20]Ibid., 1:442.

5

Paying Attention

Holiness in the Life Writings of Early Methodist Women

JOYCE QUIRING ERICKSON

I began to read the life writings—conversion narratives, diaries, letters, memoirs—of women in the Wesleyan movement in eighteenth-century Britain not for spiritual but for scholarly reasons: to explore the connections between these writings and the development of the novel as a popular genre of the eighteenth-century, connections suggested by literary and cultural historians of eighteenth-century Britain.[1] I reasoned that a reader like me who is sympathetic to the beliefs of these writers might see connections others could miss. In addition, though study of women's autobiographies has been a growth industry in literary criticism, scholars had not yet mined this particular ore. But as I sat in the library reading the yellowed pages of nineteenth-century reprints and microfilm photocopies several summers ago, I discovered my reading was more than, shall we say, aca-

[1]See, for example, J. Paul Hunter, *Before Novels: The Cultural Contexts of Eighteenth Century English Fiction* (New York: Norton, 1990).

demic. The experiences of faith chronicled in these writings touched me, despite the clichéd pious language analogous to my grandparents' religious speech that I had deliberately eschewed.

As I read these life writings, I heard echoes of these voices from the past in the agonies of a friend dying from cancer as she, like they, longed to live and die in full awareness of God's presence. I began to understand that this desire for holiness—entire sanctification—was more than a doctrinal "distinctive" that set Methodism apart from other revival movements; rather it came alive to me as a genuine desire to live deliberately in the fullness of God's loving presence. This is a goal for Christian life that contemporary Christians seldom seek or even know how to articulate, even those who come from Wesleyan or holiness traditions, much less those from traditions that emphasize the experience of justification as the primary event in a Christian's pilgrimage.

Yet the life writings of the early Methodists provide a phenomenology of holiness that is both instructive and cautionary for contemporary Christians seeking to make sense of how we might live holy lives or draw the contours of the communal life that would sustain such lives. Looking again at this experience from their perspective may clear some of the debris that has accumulated in two centuries of pious practice and discourse. In this essay I will consider as characteristic examples of this spiritual discipline the life writings of three women who were part of the intimate circle around John Wesley, acknowledged to be "mothers in Israel" by their contemporaries and descendants: Hester Roe Rogers, Mary Bosanquet Fletcher and Elizabeth Ritchie Mortimer.

The significance of band and class meetings for generating and building lives of holiness in the early Wesleyan movement is widely recognized among Methodist theologians and historians as a demonstration of John Wesley's assertion that all holiness is social holiness. But the ubiquitous life writings of early Methodists may be nearly as significant as bands and classes for building community in its testing of and testifying to the presence of perfect love. In 1781 *Arminian Magazine* announced an "addition [for] . . . the serious Reader": "part of the Life of some of those real Christians, who, having faithfully served God in their generation, have lately finished their course with joy."[2] Previous issues had included accounts of the lives of historic Christian figures such as Martin Luther; Wesley's addition of the lives of ordinary people must have reinforced his frequent instruction to practice this form of spiritual discipline.

The use of spiritual life writing as a means of spiritual discipline had been widely practiced by nonconformist groups in England for over a century, as works like John Bunyan's widely read *Grace Abounding* illustrate. Wesleyan accounts share characteristic and identifiable conventions with conversion narratives and diary accounts of writers from a variety of times and traditions. They begin with conventional descriptions of date and location of their birth and family of origin, and they exhibit a pattern of identifiable stages in the writers' experience: life before conversion, awareness and conviction of the need for salvation, acceptance of God's power to provide salvation or justification, changes in attitude and behavior that result from this assurance, and the struggles and victories in the battle to maintain the sense of assurance of the initial conversion experience.[3]

But Wesleyan accounts depart from these conventions with a significant addition. For Wesleyan Methodists not just conviction and new birth but sanctification was a "critical point of spiritual transition."[4] Thus, in all forms of life writing, early Wesleyans recounted their progress in achieving perfect love. As a result these spiritual life writings were, like band and class meetings, a significant instrumentality for substantiating holiness as the defining mark of their community. In addition to their publication in *Arminian Magazine* they were circulated in manuscript and published in monograph volumes that were frequently reprinted in Great Britain and North America.

---

[2]*Arminian Magazine* 4 (1781): 9.

[3]Virginia Lieson Brereton, *From Sin to Salvation: Stories of Women's Conversion, 1800 to the Present* (Bloomington: Indiana University Press, 1991), p. 6.

[4]Thomas R. Albin, "An Empirical Study of Early Methodist Spirituality," in *Wesleyan Theology Today: A Bicentennial Theological Consultation*, ed. Theodore Runyon (Nashville: Kingswood Books, 1985), p. 277.

Writing was a pervasive means of personal spiritual discipline. Diaries were a medium of intrapersonal communication in which writers were their own interlocutors, examining their attitudes and conduct as well as recording their blessings and shortcomings—in short, describing and diagnosing the state of their spiritual health. The anniversaries of their birth provided the occasion for a review of the past year, but even more important for such review was marking the anniversaries of their rebirth and their initial experience of sanctification. Self-examination on these dates served to test the validity of the initial experience and to remind them of the fruits that experience was expected to bear.

Reading one's diary seems to have been nearly as important as *writing* it. Their authors frequently reread them as a means of maintaining personal accountability: "A thought struck my mind to-night, as I was looking over some part of my diary, that there is not praise enough for spiritual blessings," says Mary Fletcher.[5] And later she writes:

> I have been reading over some of my old diary, and found it much blessed to me. It brought to my mind many past scenes, which increased faith and thankfulness; also, it cast a clearer light on my present state. Comparing my present state with that I felt at Hoxton, I can truly say, now I not only feel all the purity, all the spiritual mindedness, and all the resignation I did then, but in many things I prefer my present dispensation to that. Yet my soul is not satisfied, for I see a far greater salvation before me. In short, it is not the gift, but the full possession of the Giver, my spirit longs for.[6]

Elizabeth Ritchie echoes these sentiments: "I was yesterday filled with thankfulness, while reading my former journal, to feel the difference in my present experience from what it was at that time."[7]

---

[5]Henry Moore, ed., *The Life of Mrs. Mary Fletcher Compiled from Her Journal and Other Authentic Documents* (New York: G. Lane & C. B. Tippett, 1846; preface, 1817), p. 150.

[6]Ibid., pp. 236-37.

[7]Agnes Bulmer, ed., *Memoirs of Mrs. Elizabeth Mortimer: With Selections from Her Correspondence* (New York: T. Mason and G. Lane, 1836; dedication, 1835), p. 91.

Though these life writings were personal, they were not private. Reading aloud to others one's own or others' life writings was a means of encouragement, even of evangelism. An excerpt from the diary of Mrs. Bathsheba Hall from *Arminian Magazine* recounts her reading aloud to a visitor Hall's written account of her own experience of sanctification: "*While I was reading* [emphasis added], I felt the mighty power of God descend upon me, and my soul overflowed with gratitude to him, who reigned in me without a rival."[8] As this entry suggests, the act of reading aloud generated an additional benefit; it also carried out Wesley's instructions that the experience of sanctification should be told to other believers.

Further, they copied in their diaries the letters and diary excerpts of others. For example, the memoirs of William Branwell, a preacher in the Hull circuit, includes an excerpt from Hester Roe's diary in which she recounts John Fletcher's description of his experience of sanctification. Elizabeth Mortimer's memoirist includes a letter from Charles Wesley to Mary Fletcher found in Mortimer's papers.

These practices demonstrate the wide distribution of life writings among early Methodists. They also demonstrate Methodist participation in and creation of an increasingly print-oriented society, notably the participation of middle class women as writers and readers. Thus, the vexed relationship between Methodists and the dominant culture was not always adversarial; or perhaps it is more accurate to say that despite the adversarial relationship, Methodists appropriated some cultural practices unconsciously, an issue that I will take up later.

None of the women under consideration here was from the lower classes. They share other similarities. Each had a personal friendship and tie with John Wesley; they were part of an inner circle of Methodists that traveled to centers of Methodism and met in each other's homes. Each was a successful band and class leader. (Fletcher was also a preacher.) Each attained prominence and exercised leadership in the movement before her marriage.

But their life circumstances were also different, as are the forms of

---

[8]*Arminian Magazine* 4 (1781): 197.

the life writings in which their spiritual experiences are recorded. Hester Roe Rogers (1756-1794), whose life on earth was the briefest of the three, became a Methodist despite the objections of her parents. To convince her mother that she was an obedient daughter even though she refused to give up her connections with the Methodists, she became her mother's servant for nearly a year until her health was endangered. She began her ministry at nineteen and did not marry until she was twenty-seven. Present at the deathbed of the first Mrs. Rogers (who herself suggested that Hester Roe marry her husband), she became stepmother to two children and gave birth to six more, all the while continuing her ministry alongside her husband's in Ireland and then London. Her poor health prevented her from becoming John Wesley's housekeeper as he had hoped. She died from complications in childbirth. *An Account of the Experience of Hester Ann Rogers: and Her Funeral Sermon . . . , To Which Is Added Her Spiritual Letters [1796]* was edited by her husband. Her own words, whether from her diaries or letters, predominate in this volume, even in long quotations in the funeral sermon and her husband's account of her death. Since its initial publication this work has been reprinted in uncounted editions, and a reprint is even now in the catalog of a small holiness publisher.

Mary Bosanquet Fletcher (1739-1815), like Hester Rogers, became a Methodist against her parents' wishes, leaving her home to lodge in two bare rooms at their request because they feared she would influence her brothers to become Methodists too. With an inheritance from a beloved and pious grandparent she established a home for orphans and other indigents at Laytonstone, an establishment that John Wesley praised as superior to the school he had established at Kingswood. Concerned that expenses would continue to erode the capital of her inheritance, she moved to a farm in Yorkshire where she continued to maintain an establishment for the disadvantaged and disabled and to carry out her teaching and preaching ministry. At forty-two she married John Fletcher, a prominent preacher, theologian and heir-apparent to Wesley. After her husband's death only three years later she continued her ministry in his parish at Madeley until her death. *The Life of Mrs. Mary Fletcher . . . Compiled from Her Journal and Other Authentic Documents [1817]* was edited by Henry Moore (a prominent minister) to whom Fletcher had given control of her papers and diary. Nearly two-thirds of this volume are excerpts from her diary after John Fletcher's death; thus, her reflections are not those of a young Christian but of one who is fully acculturated into the community. The importance of Mary Fletcher in the life of the Wesleyan community is perhaps evidenced by the inclusion of these memoirs as one of the volumes in her husband's collected works.

Elizabeth Ritchie Mortimer (1754-1835) was the woman chosen to become housekeeper to Wesley when Hester Rogers could not. Born into a pious family, in her adolescence she was ashamed of their Methodist connections, preferring the more worldly life offered to her by a wealthy godmother. By age seventeen, however, she became convinced of the truth of her family's faith, partly through the ministry of John Wesley during visits to her family home. She often traveled with Wesley and led band and class meetings. She refused an offer of marriage from John Mortimer, but when he asked her again years later, she accepted and was married at age forty-seven, becoming the beloved stepmother of two daughters. *Memoirs of Mrs. Elizabeth Mortimer: With Selections from Her Correspondence [1836]*, edited by Agnes Bulmer, was published the year after her death. Bulmer's narrative summaries and pious commentary predominate in the memoir, and her selections from Mortimer's own writings tend to downplay the extent of Mortimer's ministry, undoubtedly reflecting the changing attitudes toward women's leadership in the movement as it approached its centenary. Mortimer appears to have ended the practice of writing a diary long before her death, and Bulmer's transcription of the extant diary entries leaves out some sections because, she claims, they were written for private not public benefit.

Bulmer's comment suggests that not all of the diaries were written to be read by others, though it may also be that some of Mortimer's entries were embarrassing to a movement that was becoming more respectable. In each of these works, in fact, the reader's access is mediated by an editor whose own aims guide the selection.[9] Ironically, this mediation supports the claim that these life writings are communal documents. As acknowledged community leaders, the editors presumably selected material that would reinforce or shape community life.

What are the significant experiences in the recorded lives of these three women? Particular similarities lead to several conclusions: (1) the initial experience of sanctification is preceded by the same kind of conviction of need and emotional distress that precedes the experience of justification; (2) the experience of sanctification is not accompanied by an emotional high (as is sometimes the case with the experience of justification); and (3) the experience does not alleviate the writer's struggles to be faithful.

Each writer is careful to note the circumstance, time and date of the initial experience of sanctification. This experience comes after months or years of expressed desire for the experience, accompanied by a concern that a counterfeit experience might be mistaken for the real thing. The quest for sanctification these writers describe is neither solitary nor devoid of theological reflection. The writers recount their participation in all of the means of grace prevalent in early Methodism: spiritual conversation, participation in bands and classes, attendance at the public reading and exposition of Scripture, receiving the sacrament of Communion, and theological reading, particularly John Wesley's sermons and essays. The self-examination recorded in their diaries as they struggle to receive perfect love is thus only one aspect of an experience in a community rich with spiritual resources.

These resources are, in the words of Mary Bosanquet as she describes her search for full salvation,

> rivers of living water [that] flow from heart to heart. . . . Some portion of this river seemed now to reach me also. The means of grace were as

[9]Rogers's work is the least edited, which, along with Moore's and Bulmer's occasional editorial comments explaining or correcting the theology or practice of the writer, supports the thesis that editors' choices were guided by a sense of their audience as the movement became more acclimated to the surrounding culture.

> marrow to my soul. . . . But I could not believe so as to give up my whole heart to the Lord. I knew him mine, but other things had yet life in me, though not dominion over me. I was now assured the blessing of sanctification (or, in other words, a heart entirely renewed) could not be received but by simple, naked faith; and my soul groaned out its desire.[10]

Hester Roe read and reread Wesley's *Plain Account* and *Further Thoughts on Christian Perfection* (as her experience of justification had been preceded by reading Wesley's sermon on justification by faith), as well as works by John Fletcher.

> I now was powerfully convinced, that whenever sin is totally destroyed, it is done in a moment. . . . I had a deeper sense of my impurity than ever; and though by grace I was restrained from giving way outwardly, yet I felt such inward impatience, pride, fretfulness, and, in short, every ill temper, that at times I could truly say, I was weary and heavy-laden.[11]

> Saturday, 27 [1776]—Mr. Wesley's Plain Account of Christian Perfection was this day a greater blessing than before. . . . I find, while pressing after entire purity, my communion with God increases, and I have more power to do his will. Friday, February 2.—I awoke several times in the night, praying for sanctification. O, the depth of unbelief and of pride![12]

The diary entry after her experience of sanctification begins with the same four words but ends quite differently: "O the depth of solid peace my soul now felt!"[13] The line that follows this utterance admits she has "not so much rapturous joy as at justification," but her resolution to postpone telling others of her experience is broken when "it was seen in my countenance; and when asked respecting it, I durst not deny the wonders of his

[10]Moore, *Life of Mary Fletcher*, pp. 35-36.

[11]*Experience of Hester Ann Rogers and Her Funeral Sermon, By Rev. Dr. Coke, To Which Is Added Her Spiritual Letters* (1796; reprint, Cincinnati: Hitchcock & Walden, n.d.), p. 41.

[12]Ibid., p. 43.

[13]Ibid., p. 47.

love!"[14] In a letter to her cousin Robert with whom she carried on an extended correspondence concerning his doubts about the possibility of sanctification, she is exultant: "You ask if I am not in a delusion respecting my experience of perfect love? Blessed be God! I have not the shadow of doubt. Even Satan himself finds these suggestions vain, and has left them off. . . . You ask how I obtained this great salvation? I answer, Just as I obtained the pardon of my sin—*by simple faith*."[15]

Similar themes are evident in Mary Bosanquet's description of her experience:

> Deep discouragement seized my spirit—but I wrestled on, and was in an agony to *love God with all my heart*. Brother Gilford was praying for me, when in a moment I felt a calmness overspread my spirit, and by faith I laid hold on Jesus, as my full Saviour. . . . I found that the love of the will of God had brought an unspeakable peace into my soul: but that I did not feel joy; only a rest in that thought, *The Lord reigneth, and his will shall be done*.[16]

Such peace does not remain, however, and she interprets her subsequent discomfort as "a low degree of pure love. . . . I had many temptations, and not much joy. Yet did I never feel any thing contrary to love; and in the temptations with which I was attached, I felt a great difference."[17]

Elizabeth Ritchie, having experienced justification, also learns there is more to desire: "experience soon taught me, that my warfare was but just commenced. For some time I was all love, prayer, and praise; but painful circumstances soon convinced me, that the propensity to evil had been only dormant in my heart: wrong tempers yet remained. . . . [I] felt the need of a more full salvation, and was resolved to seek it earnestly."[18]

She began her diary on July 3, 1771 "as a help to [her] spiritual

---

[14]Ibid., p. 48.
[15]Ibid., pp. 207-8.
[16]Moore, *Life of Mary Fletcher*, p. 36.
[17]Ibid., p. 37.
[18]Bulmer, *Memoirs of Elizabeth Mortimer*, pp. 32-33.

progress . . . [to] regularly note down what passed in [her] mind, relative to this important subject."[19] Subsequent entries record her struggle: "The Lord knows my heart, and I can appeal to him that I sincerely desire to be delivered from all unholy tempers: they break my peace, and bring me into bondage."[20] At one point she asserts that "the Lord has wrought a great change in me,"[21] but an entry several months later begins, "I have made but little progress. The corruptions of my heart have been very lively."[22] It is nearly three years (May 1774) until she is able to claim "that a blessed change was effected in me."[23]

Though these writers make a careful distinction between their feelings and their spiritual state, they do not therefore pay less attention to their emotional state. This close and continual scrutiny of their attitudes and conduct is so integral to their sense of progressing in holiness that taking their spiritual temperature inevitably leads to taking their emotional temperature. A passage from Hester Rogers' diary a few years before her death is paradigmatic of the lifelong struggle. She attributes the "cloud of heaviness" that has "at some seasons, hung upon my mind" to difficult circumstances in her husband's ministry:

> Satan has taken occasion to suggest, in those times of depression, various accusations of shortcomings in zeal, activity, and spiritual joy. I do not mean that I was ever left in darkness; no, since I first consciously received a sense of favor with God, I never lost it; but within two years last past, I have not always had so clear a witness of perfect love. At other times I have had that witness full and clear; . . . But in nothing else than full salvation, and the witness of it, could my soul ever rest.[24]

A letter to her cousin written years earlier had predicted this possibil-

[19]Ibid., p. 33.
[20]Ibid., p. 38.
[21]Ibid., p. 41.
[22]Ibid., p. 42.
[23]Ibid., p. 45.
[24]*Experience of Hester Ann Rogers*, p. 77.

ity: "Our joy may at times be small; yet our faith may be perfect, and our peace undisturbed."[25]

It is also not surprising, given the many years Mary Fletcher kept a spiritual diary, that recurring cycles of blessed happiness and depression characterize her experience as well. As a newly sanctified young woman she marvels at her extraordinary awareness of her emotional state:

> It seemed most exquisite feelings were opened in my soul, such as I never knew before. If I saw or heard of the consequences of sin, I was ready to die! For instance,—if in the street I saw a child ill used or slighted by the person who seemed to have the care of it, or a poor person sweating under an uncommonly heavy burden; or if I saw a horse, or a dog, oppressed or wounded, it was more than I could bear. I seemed to groan and travail in birth, as it were, for the whole creation. Yet notwithstanding all these painful feelings, I had a solid peace.[26]

Years later, the emotional devastation she felt at her husband's death was of great concern to her: "If Jesus was my all, should I not feel as keenly the sense of his having suffered for me, as I do in the thought of my dear husband's kindness, and in the dreadful feeling of my separation from Him? And because I could feel but very faint touches of sensible communion with God, I was torn as it were in pieces."[27] She acknowledges she is "not to set joy as the mark,"[28] but diary entries over the years often lament the absence of "sensible joy." Humility and resignation to God's will provides the only hope for relief from "despairing thoughts, sent by the devil."[29]

Though her diary records her frequent sense of inadequacy and though she never ceases to examine with incisive honesty her attitudes, motives and conducts, as years pass her affirmations of God's presence and her willingness to abide in his will do predominate. "I

---

[25]Ibid., p. 225.
[26]Moore, *Life of Mary Fletcher*, pp. 37-38.
[27]Ibid., p. 165.
[28]Ibid., p. 208.
[29]Ibid., p. 245.

am sure I do feel an increasing resignation, and that not in theory, but in practice."[30] Not in theory, but in practice. For all of these writers it is practice—action, including the action of internal reflection—that provides the significant evidence of perfect love. One might say, though admittedly Mary Fletcher would not put it quite this way, that feelings are equivalent to theory.

It is either an extremely sensitive or an extremely crude emotional register that distinguishes between states of joy, peace, and happiness. In our contemporary descriptions of spiritual health, these terms are flattened into synonyms, even empty clichés. But most present-day users of such terms are not engaged in the sort of theological reflection, Scripture study and regular participation in other means of grace (preaching, holy Communion) that were part of the rhythm of these women's lives. In that context Hester Rogers's impassioned exclamation at the end of the despairing passage quoted above is poignant but also puzzling: "O no! What is past experience without present enjoyment? I must feel, or I can not be happy."[31] Despite all protestations to the contrary she and her contemporaries find it impossible to separate their apprehension of perfect love from their feelings and emotions.

Perhaps there is something about the very medium of life writing, especially the daily spiritual journal, that fosters a misrepresentation of experience since its purpose is not to record the minutiae of the daily round as many diarists do but to chronicle the minutiae of spiritual life. Conspicuously absent from these journals are the domestic details that surely dominated their lives, as the following summary of life at Laytonstone by Mary Bosanquet shows:

> We sometimes had much to do; for the care of the sick, the management of eighteen or twenty children, with various meetings, and the needful attention to the work of God in a new raised society; with the reception of the number of strangers who visited us on spiritual accounts, occasioned those of us who had the work of God at heart, a

---

[30]Ibid., p. 306.
[31]Ibid., p. 77.

good deal of labour and suffering.[32]

References to public affairs or political concerns in these diaries are rare, unless mentioned as subjects of prayer (e.g., a threatened invasion of the French during the Napoleonic wars). In an uncommon mention of the pleasures of the natural world Elizabeth Mortimer follows this up with an exhortation that minimizes that experience: "You and I, my dear friend, have often seen immortal spirits vainly striving to quench their thirst for happiness at the streams of creature comfort. Disappointment has been the result."[33] And with respect to marriage, an event that would dominate conventional diaries, Hester Rogers only suggests that the details leading to her marriage are "too tedious to dwell upon here." Their marriage is summarized by the affirmation that in Mr. Rogers "the Lord gave me a helpmeet indeed; just such a partner as my weakness needed to strengthen me."[34]

The salient point is that every external event recorded in these diaries is contextualized by a larger spiritual claim, as Rogers's description of her marriage illustrates. This is true as well of the one aspect of ordinary life that does persist in these diaries—close friendships with other women. The connection with other women of faith whose support—material and spiritual—provides the strength to go on is a dominant theme in the diaries and letters. Without these "holy friendships," their writers implicitly and explicitly suggest, living holy lives would be nearly inconceivable. Such friendships are part of the communal network that sustained them, including John Wesley himself, as his letters attest.

Wesley's discussion of his own experience in a letter to Elizabeth Ritchie (included in her memoirs) gives us a hint about how we might conceptualize the early Methodists' understanding of the connection between the holy life and feeling.

I do not remember to have heard or read any thing like my own expe-

---

[32]Moore, *Life of Mary Fletcher*, p. 51.
[33]Bulmer, *Memoirs of Elizabeth Mortimer*, p. 247.
[34]*Experience of Hester Ann Rogers*, p. 72.

rience. Almost ever since I can remember, I have been led in a peculiar way. I go on in an even line, being very little raised at one time, or depressed at another. Count Zinzendorf observes, "There are three different ways wherein it pleases God to lead his people. Some are guided, almost in every instance, by apposite texts of Scripture; others see a clear and plain reason for every thing they are to do; and yet others are led, not so much by Scripture and reason, as by particular impressions." I am very rarely led by impressions, but generally by reason and Scripture: I *see* abundantly more than I *feel*. I want to feel more love and zeal for God.[35]

Wesley's expressed desire to *feel* more is not matched by the converse, the expression of a desire to *reason* more by those for whom feeling is prevalent. Consider Mary Fletcher's approbation of the following sentiment expressed in a letter from a friend: "When we look at Jesus by faith, Satan loses his power, and, if I may so speak, his place, which is the reasoning faculty."[36] Or this advice to a friend by Elizabeth Ritchie:

> Your mind rises into what is rational; but I want you to enjoy what is spiritual: pray for power. At various times the Holy Spirit has graciously touched your heart with a sense of want; but your studies have proved to this blessed spark in you, what the foolish trifles of a moment have been in many others;—they engross your mind, so that its entire vigour is drawn into them. Let me entreat you to be determined that you will take time for private prayer, and for reading the Scriptures as the revealed will of God. *They will help you to read yourself;* . . . The soul that sees itself will earnestly desire Christ.[37]

Though such comments are not sufficient to charge Fletcher or Ritchie with advocating irrationality, they do suggest that of Wesley's three ways of apprehension, feeling is preferred to reason, though as reason is guided by Scripture so is feeling.[38]

---

[35]Bulmer, *Memoirs of Elizabeth Mortimer*, p. 97.

[36]Moore, *Life of Mary Fletcher*, p. 313.

[37]Bulmer, *Memoirs of Elizabeth Mortimer*, p. 150 (emphasis added).

[38]Perhaps one should not make too much of statements in personal letters not intended as carefully argued reflections about human nature of the kind discussed in

I began by making the claim that examining the phenomenology of holiness in the life writings of these eighteenth-century forebears would provide clues for twenty-first century Christians who wish to lead holy lives. The most striking observation to be made in comparing their practice with ours is the degree of seriousness with which they pursued a holy life. Their disciplined lives look much less like the peasants celebrated by Stanley Hauerwas than like Christians in monastic communities. They assiduously practiced the means of grace that would foster holy living, and they did it as part of a community. Even the personal and private act of keeping a spiritual journal became a communal vehicle for cultivating "holy tempers." When we compare the means of grace by which the early Wesleyans sustained holiness and community, both in quantity and quality, to the meager fare with which most contemporary Christians are satisfied, our spiritual atrophy is not surprising.

The phrase that best characterizes their stance is "paying attention." Most of us, even committed Christians, pay precious little attention to the ways our large and small actions affect our spiritual condition. But to determine how the stance of paying attention might look in the twenty-first century requires a more careful consideration of the significant differences between the conditions that characterized eighteenth-century life and our own.

The daily lives of eighteenth-century Europeans were dominated by life's limitations in a relatively small universe of possibilities. Twenty-first century people are presented with nearly too many choices in a vast universe of possibilities (though the poor in both prosperous and developing nations are closer to the condition of the eighteenth-century than to ours). Nowhere is that difference in limitations more evident than in the different conditions of physical health in the eighteenth and twentieth centuries. The everyday reality of death in earlier centuries is virtually incomprehensible to us: a healthy person could catch what appeared to be a cold and be dead

---

Theodore Runyon's essay in this volume (chapter four). There is no doubt that these writers accepted implicitly Wesley's notion of "spiritual respiration" as a description of holiness.

in three days. The outcome of pregnancy and childbirth could never be predicted. Medicine was in its infancy as a science. And without aspirin or other anodynes that we take for granted, the aches and pains endemic even to healthy people were pervasive. It is no wonder that diaries and letters frequently note their physical illness or discomfort (a characteristic of all eighteenth- and nineteenth-century life writings) or that these early Wesleyans saw a clear link between spiritual and physical health. Their stories indeed exemplifed the "testimony" that Stanley Hauerwas calls "the communicative body" (p. 32).

Their experience led them to assume an interconnection between physical and spiritual health, connections that contemporary practice is only beginning to acknowledge. Illness reminded them of their dependence on God and provided an occasion to strengthen their relationship with God, as Wesley implies in a letter to Elizabeth Ritchie: "It is an admirable providence which keeps you thus weak in body, till your soul has received more strength."[39] Wesley's promotion of medical remedies and habits for maintaining good health forbids a misreading of this statement as a lack of concern for physical well being. Rather, it is an acknowledgment that the physical and spiritual are not separable; in Hauerwas's terms it is an opportunity to bear witness through the body (p. 35).

The everyday occurrences of illness and death fostered a consciousness of death's inevitability but also some ambiguity, since the next world was seen as infinitely preferable to this one. The young Hester Roe, who seemed likely to die from "a consumption," refused to take medication in hopes that she would be able to meet her Savior in heaven. She records actual disappointment when she appears to be recovering, though she repents when her cousin chides her for setting her will against God's.[40] Each of the women considered here experienced life-threatening illnesses that brought her close to death. They interpreted their recovery less as a reprieve from disaster than as a clear sign that there was more to do in this

[39]Bulmer, *Memoirs of Elizabeth Mortimer*, p. 55.
[40]*Experience of Hester Ann Rogers*, p. 39.

world. Though people who appeared close to death were treated with all available means to effect a cure, physicians or loved ones were not faced with anguished questions about whether life should be prolonged. At any rate, the next world was a better place to be if one were right with God; death was a triumph for the sanctified Christian, and the frequent deathbed scenes recounted in spiritual life writings record that triumph as a witness for unbelievers and backsliders.

Although some contemporary Christians may occasionally express a conviction that the next life is preferable to this one (more perhaps in developing countries than in prosperous Europe or North America), it is rarely a topic of conversation or a prevailing motif in hymns as it was in the eighteenth and nineteenth centuries. Had the early Methodists been given a twentieth-century diagnostic test for depression, they would surely mark this phrase: "I often think about dying." If we marked that phrase, we would meet our health insurance eligibility criteria for treatment. To refuse treatment or to consider the possibility that our emotional distress might be a sign of spiritual distress would be the equivalent of Hester Roe's refusal of treatment for her consumption. Even if we acknowledge that illness does indeed often affect our spiritual lives, few of us would commend illness as a means for achieving spiritual maturity or living holy lives. In our therapeutic society the means to relieve our physical and emotional illnesses are readily available. Yet our dependence on Prozac or other nostrums ironically means we are deprived of one of the means of progressing in holiness available to the early Wesleyans. As Hauerwas says, "Illness may be the only time that we have the opportunity to discover that we are part of a story that we did not make up" (p. 29). These early Methodists recognized in physical or emotional distress the signals of potential spiritual danger that requires more than medical treatment for relief.

Having acknowledged this, there is still a question whether the depression and low spirits frequently reported in these life writings may have been exacerbated by the intense self-examination that was deemed a necessary means for maintaining sanctification. Anti-

Methodist writers of the eighteenth century claimed that Methodists, particularly Methodist women, were susceptible to the "disease of melancholy."[41] Paradoxically, achieving sanctification *increases* rather than diminishes self-investigation and reliance on "impressions" (the word Wesley associates with feeling in the letter quoted above) to gauge one's spiritual health. Any moderately self-reflective person will find it frequently difficult to maintain a sense of holiness (whereas just the opposite might be said about any moderately self-righteous person).

Escaping the difficulties and pains of this life, however, was not the primary reason for denying this world in preference for the next. The writers of these diaries preferred the next world because in that world they would be face to face with Jesus. The ultimate goal of holiness, as so many of Charles Wesley's hymns remind us, is *theosis,* to be "lost in wonder, love and grace." Not only did the pleasures offered by this world pale in comparison to that expected experience, it also hindered present fellowship with Jesus and threw up distractions that were impediments to holiness.

Renouncing worldly pleasures is a dominant theme in discourses on holy living from all traditions, and it is a *leitmotif* in Methodist writings. Giving up many of the pleasures that their contemporaries sought was the first action Methodists took as they underwent the experience of repentance. Mary Bosanquet's father is incredulous that she would consider to be sinful "all places of diversion, all dress and company, nay all agreeable liveliness, and the whole spirit of the world." But she asserts the disjunction unequivocally: *"The friendship of this world is enmity with God."*[42] Embarking on the path of salvation requires a conscious choice, in the words of Elizabeth Ritchie, between "shadow or substance,—Christ or the world,—profession only and self-pleasing, or the possession of a present Saviour, and that self-denying path which, in lines marked

[41]Cynthia J. Cupples, "Pious Ladies and Methodist Madams: Sex and Gender in Anti-Methodist Writings of Eighteenth Century England," *Princeton Working Papers in Women's Studies* 5 (spring-summer 1990): 55.

[42]Moore, *Life of Mary Fletcher,* p. 22 (emphasis in original).

with blood Divine, leads to eternal glory."[43]

Few such dramatic sacrifices are expected of late twentieth-century Christians, as Hauerwas and others have frequently reminded us. In the face of the rampant hedonism and consumerism that characterizes Western culture, it would seem we have much to learn from these early Methodist forebears, as well as from those in the American holiness churches who have equated a *world-denying* stance with a faithful expression of the desire for, if not the manifestation of, holiness and perfect love. Yet many of us are familiar with the perversions of a world-denying stance turned into a *life-denying* stance which ultimately only substitutes fashionable aspects of material culture for the unfashionable, the elegant for the dowdy or shoddy. This strain of asceticism is one of the most problematic aspects in seeking holiness for late twentieth-century Christians, certainly for professors in liberal arts institutions who consider it part of their vocation to help students appreciate the arts or literature as a gift from God!

What aspects of self-denial are necessary (though not sufficient) for holiness? Are the pleasures of the material world and of human culture inevitably destructive of faithfulness? Not all Christian traditions have thought so, of course. And even as the early Methodists eschewed the blandishments of their contemporary culture, they were unwittingly responding to and incorporating eighteenth-century proclivities into their own practice. The affinities between the literature of sensibility and sentiment that flourished in the eighteenth century and Wesleyan emotionalism is nearly a commonplace of literary history. Mary Fletcher's description of her anguish at seeing an animal suffer quoted earlier could serve as a prime example of the appropriate response of a man or woman of "sensibility" to such phenomena, since susceptibility to and expression of emotional anguish provided evidence that one possessed "sensibility." John Wesley himself edited and published one of the most popular of the novels of sensibility, Henry Moreland's *The Fool of Quality.* (Wesley disliked the title, however, so he changed it to *The History of Henry Moreland.*) Similar

---

[43]Ibid., p. 148.

observations could be made about the uses of life writing outside of the Methodist community. Even the theater, suggests Henry Abelove, provides an analogy to Methodist practice:

> They had an ongoing theater of their own, which they liked better than the one that dramatists provided. In their theater they were the stars as well as the audience. Their lines were the lines that were remembered and commented upon afterward. Their concerns were the subject of the play, and furthermore they needed to make no rare and hard effort of sensibility to grasp what was happening around them. For the words were familiar, even if heightened, and the others present were just their usual companions, even if especially excited for the occasion.[44]

Abelove may merely be the most recent in a long line of unsympathetic outsiders who caricature Methodist practice. And we perhaps need pay no more attention to this opinion than the early Methodists paid to the torrent of vitriolic and demeaning satire. Yet the question remains whether *some* versions of the so-called good life are inexorably antithetical to the holy life.

The Methodists' resistance to the cultural norms of their time is positive as well as negative. At its most positive it is inclusiveness that overrides class and gender distinctions, affirming that the good life of salvation is available to everyone. And it also affirms the power of the Holy Spirit to overcome and subvert the prevailing social structures, evident in the public roles women enacted as band and class leaders and as preachers. This was not an easy role for the women, since they often did so in the face of objections from some within the movement and ridicule and slanderous accusations from without. Mary Bosanquet's diary records these difficulties:

> This day I set apart for prayer, to inquire of the Lord, why I am so held in bondage about speaking in public. It cannot be expressed what I suffer—it is known only to God what trials I go through in that respect. Lord, give me more humility, and then I shall not care for anything but thee![45]

---

[44]Henry Abelove, *The Evangelist of Desire: John Wesley and the Methodists* (Stanford, Calif.: Stanford University Press, 1990), p. 105.
[45]Ibid., p. 103.

Public humiliation was one of the obvious difficulties women experienced in maintaining their leadership. Carrying out conventional gender roles as well as unconventional roles was less obvious but no less difficult. (Recall the summary of Bosanquet's domestic responsibilities quoted earlier.)

This difficulty is illustrated in Elizabeth Ritchie's allusion to the biblical story of Mary and Martha in which Mary "chose the better part" to listen at Jesus' feet. She advises a newly married friend that despite her "change of situation," the friend's call is "to Mary's situation." Ritchie is concerned that the "outward employments" her friend must now take up might "hinder close attention to the voice of God."[46] Ritchie herself assures John Wesley in a letter that "when I am called to serve with Martha's hands, I feel a Mary's heart."[47] On the one hand the story of Mary and Martha validates women's spiritual equality with men and affirms their participation in the life of the community as more important than their domestic contributions. On the other hand it places them in the classic psychological double bind, because the practical reality of their lives requires their participation in all of the expected domestic routines that distracted them from devotional activity and took up their waking hours. Since responsiveness to the material and spiritual needs of others is a mark of holy living, the pressure is increased.

In fact, except for the exercise of spiritual leadership in public, the early Methodists did not challenge conventional gender roles. The hierarchical order of wifely submission to her husband is assumed. John Fletcher on his deathbed bequeaths his headship over Mary to Christ.[48] Dr. Coke's funeral sermon praises Hester Rogers for her ability to fulfill her many roles so admirably without slighting her maternal duties, and her husband obliquely praises her for *not* preaching in public: "Notwithstanding her extraordinary zeal for God and the salvation of souls, her good sense, joined with that Christian modesty which is ever becoming her sex, taught her as to the manner how to

---

[46]Bulmer, *Memoirs of Elizabeth Mortimer*, p. 64.
[47]Ibid., p. 95.
[48]Moore, *Life of Mary Fletcher*, p. 161.

proceed in saving souls from death."[49] Bulmer's account of Mortimer praises her private ministry but downplays her prominence as a public leader.

By the early nineteenth century few women were preaching. In the dominant culture the doctrine of separate spheres for men and women was gaining ascendancy, denying women access to action in the public sphere but giving them moral dominance in the private domestic sphere. This "doctrine" led to an increasing split between private and public and to an ascription of gender-specific roles for both men and women in the nineteenth century that exceeded gender restrictions of the eighteenth century, including the pernicious assumption of women's moral superiority, a superiority they were able to maintain only because they did not soil themselves in the dirty arenas of public life. Methodists did not challenge these assumptions. I suspect that the spiritualizing discourse of Methodist women's life writing reinforced this ideology for later readers.

The separation of private and public life is conducive neither to holy living nor to healthy community. "Public and private are two different things," responded one citizen (echoing Richard Nixon two decades earlier) on a recent National Public Radio interview about White House sex scandals, and Wesley's twentieth-century descendants seem to have agreed, equating holy living with one or the other: personal or social morality, private or public faithfulness to the imperatives of the gospel. But both the public and private spheres must be seen as a locus of significant human activity and source of meaning for human life, argues Christian political theorist Jean Bethke Elshtain as she calls for "the redemption of everyday life."[50] It is significant that in the Methodist women's life writings considered here, neither public nor private sphere emerges as visible. The public sphere is attenuated and apolitical, since the women (like most eighteenth-century women) are effectively excluded—or exclude themselves—from participation in a wider public cultural

---

[49] *Experience of Hester Ann Rogers,* p. 133.

[50] Jean Bethke Elshtain, *Public Man, Private Woman: Women in Social and Political Thought* (Princeton, N.J.: Princeton University Press, 1981), p. 335.

conversation about political or cultural life. "Everyday life" in the private sphere is nearly as invisible as the public sphere. Their spiritual language, whether private or uttered in community, obliterates the discourse of public or private life. For women this means giving up their voice to patriarchal authorities in the private sphere, and for believers it means giving up their voice to secular authorities in the public sphere.

To me the dismissal of everyday life in these writings that is most troubling is the absolute absence of humor. Holy joy may include spiritual ecstasy, but it does not appear to include laughter. (One recalls that the word *serious* was virtually synonymous with *evangelical* as a descriptor of Christians in the nineteenth century.) Perhaps it is not surprising that these Methodists found it difficult to laugh; as the objects of satire more brutal than amusing they were understandably not amused. There may have been humor and joyful laughter in their daily lives that they did not record because it was perceived to be irrelevant to the intense and concentrated investigation of their spiritual condition, just as there may have been pleasure and comfort in the sensuality of married love that is also not recorded, but if we seek in them models for contemporary ways of pursuing holiness, these are telling omissions.

Nevertheless, we have much to learn much from their experience and practice. Certainly they knew how to pay attention to what they believed mattered most: living lives that expressed the perfect love of Christ. The disciplines they followed kept this goal foremost in their consciousness. From them we can learn that the *desire* for sanctification precedes its advent, a desire that is rarely expressed or cultivated among contemporary Christians. And that desire was accompanied by assiduously seeking the means of grace that brought sanctification and holy tempers: prayer, study of the Scriptures, band and class meetings, public meetings, the sacrament of holy Communion, fasting, spiritual friendships. All of these means are signals that kept them paying attention, like the signals on a train track that remind the engineer he or she is nearing the destination.

Notwithstanding the intense self-examination represented in these life writings, the aim was not self-knowledge but knowledge of Christ. The currency of their pervasive pietistic language has been debased, but the reader who accepts their premise will soon recognize genuine coin. And it is no more debased than the psychobabble that pervades much popular religious discourse. They do not write to boost their self-esteem but to find God. Our difficulty in appropriating their practice is that every one of the means they used has been appropriated by popular culture as a means of "self-fulfillment."

But let us suppose. Suppose we as contemporary Christians learned to desire holiness and learned from our forebears how to fix our attention on that desire—to *pay attention.* In our media-saturated society we may need to first learn how to fix our attention sufficiently to benefit from attention. But once our attention is fixed, we may, like our forebears, use the private and communal disciplines already available to us. Let us suppose also, however, that we add to this practice additional signals that are not noted in their writings, markers from the minutiae of our daily lives that trigger our reflection. Suppose we sought the means of grace in holy Communion as frequently as they did, but suppose we also named the food and drink of our daily lives a sacrament. Suppose we wrote a journal that recorded not our feelings but the *actions* we have carried out or witnessed that reflect—or deflect—love. Suppose we noted in that journal the state of our physical health or illness, deliberately linking our bodily life to our mind and spirit. Suppose we made a covenant with a spiritual friend to speak honestly to one another about how these disciplines have actuated or diminished our caring for the least of the kingdom.

Then we might be taking the best from their model. None of these suggested activities is novel in the panoply of spiritual disciplines currently practiced by Christians from a variety of traditions. But suppose we articulated to ourselves and to each other in our Christian community that the goal of such discipline is sanctification, perfect love, holy living. Suppose that the signals we used to help us pay attention were world-affirming and life-affirming, understood as gifts of God's creation and gracious presence through which the events and objects of our daily lives are transformed into sacraments. And suppose that even as we affirmed these gifts, we were careful to acknowledge their power to distract us from the One on whom our eyes should be fixed, their power to become idols.

This kind of paying attention is action, not theory, to use Mary Fletcher's words. She and her contemporaries lived the knowledge that we are utterly dependent for our holiness, our wholeness, on the Spirit's action in our hearts. As she and her friends knew, only in community is it possible to pay such attention and to bear witness in our bodies, Christ's body.

6

The Once & Future Church *Revisited*

MICHAEL G. CARTWRIGHT

The pastor concluded his message to the congregation of "First Church" with these words, a challenge that sent them home with new resolve, a new mission:

> The call of this dying century, and of the new one soon to be, is a call for a new discipleship . . . more like the early simple, apostolic Christianity. . . . Nothing but a discipleship of this kind can face the destructive selfishness of this age with any hope of overcoming it. There is a great quantity of nominal Christianity to-day. There is need of more of the real kind. We need a revival of the Christianity of Christ.[1]

The words of this fictional sermon—written a century ago—display the nostalgic impulse to *reinvent* Christian holiness in the midst of its apparent absence. The discrepancy between nominal Christianity and "the real kind" is what generates the spiritual yearning for a future

[1]Charles Sheldon, *In His Steps*, Family Inspirational Library (New York: G.P. Putnam's Sons, 1982), p. 237. This is regarded as the "authorized" edition of Sheldon's work.

church that would replicate the apostolic church. At the same time, as Stanley Hauerwas has argued, it is precisely because the language and practice of holiness in American Protestantism is "too spiritual" that stories of the embodiment of discipleship by *actual* saints and congregations are being silenced (p. 36). Succumbing to the temptation to "start fresh without remembrance,"[2] American Protestants fantasize about "the once and future church," thereby displaying both our lack of practical wisdom and some highly questionable assumptions about ecclesiology.

Although its origins are not well understood, the saga of "the once and future church" is a familiar tale. It is a story of the decline—and the yearning for renewal and revitalization—of Christian congregations. It is a story that unfolds in the midst of confusion precipitated by unforeseen events that in turn prompt uncomfortable reflections about the present and future. According to some prominent analysts of con-

---

[2]T. H. White, *The Once and Future King* (New York: Ace Books, 1939), pp. 631-32. The full context of this quote is provided in the passage below.

> "Let us start fresh without remembrance, rather than live forward and backward at the same time. We cannot build the future by avenging the past. Let us sit down as brothers and accept the Peace of God."
>
> Unfortunately men did say this, in each successive war. They were always saying that the present one was to be the last, and afterward there was to be a heaven. They were always to rebuild such a new world as never was seen. When the time came, however, they were too stupid. They were like children crying out that they would build a house—but when it came time for building, they had not the practical ability. They did not know the way to choose the right materials.

This ironic passage from *The Once and Future King* aptly describes an American habit of mind—the desire to escape the problems of the present by seeking a future detached from the past. The moral context is that of war, but it is applicable to other moral issues as well. White's version of "Arthur/Wart" is a paradigm of disoriented practical judgment, and White's Merlyn is a figure who lives backward in time, knowing what is to come but always with a tenuous grasp on the present. Taken together these characters display the *displacement* of practical wisdom in the American cultural context.

As the broader context of the passage quoted above also makes clear, there is not just the practice of amnesia but also the longing for "the blessing of forgetfulness" (p. 631) in the face of seemingly perpetual conflict with the stranger in one's midst. Perhaps this accounts in part for the spate of books by and for American Protestants about church renewal that in different ways invoke the Arthurian imagery of Camelot.

gregational decline and renewal it is a drama about "a fundamental change in how we understand the mission of the church. Beneath the confusion we are being stretched between a great vision of the past and a new vision that is not yet fully formed."[3] As I will argue in this chapter, the fantasy of "the once and future church" can also be seen as a site for the identification and critique of the recurrent constructions of holiness and community that have emerged in the midst of the disestablishment of mainline American Protestant congregations.[4]

In this chapter I argue that in twentieth-century American Protestantism the *fantasy* of holiness has been more prominent than the practice. Or to state my claim in a philosophical mode, the *language* of holiness has become so severely disordered that we use the classic catch phrases *as if* they designate something socially embodied, when they actually do not. In sum, we find ourselves in a circumstance in which "we continue to use many of the key expressions" of holiness.[5] We have *simulacra* (empty conceptual structures or artificial imitations of a real thing) of holiness; however, we have largely if not entirely lost our comprehension, both theoretical and practical, of what it means to embody Christian holiness. Why this is the case and how things came to be the way they now are can be partially explained by exploring the text of Charles Sheldon's novel *In His Steps,* first published in 1897. In this analysis I will identify some of

---

[3]Loren B. Mead, *The Once and Future Church: Reinventing the Congregation for a New Mission Frontier* (Washington, D.C.: Alban Institute, 1991).

[4]Although I have elected to use the phrase "the once and future church" throughout this essay, in doing so I do not intend to use Loren Mead's book as the target of my critique except insofar as the series of books that he has edited displays this same penchant for "reinventing" the church without taking into account the ways in which his own prognoses for remedying the situation end up disembodying the church. To take but one example, in Mead's *Five Challenges for the Once and Future Church* (Bethesda. Md.: Alban Institute, 1996), the most recent book in the Once and Future Series lists as its "fifth challenge" to become an apostolic people (as opposed to trying to embody what it means to be an "apostolic church") who would seek to live out "God's mission." Tellingly, Mead contends that "God's distinct mission today is not necessarily directed through the church" (p. 74).

[5]This analysis is a paraphrase and transposition of a claim made by Alasdair MacIntyre in *After Virtue: An Essay in Moral Theory,* 2nd ed. (Notre Dame, Ind.: University of Notre Dame Press, 1984), pp. 1-2.

the many *simulacra* that distract us from engaging in authentic Christian practices of holiness and community.

In His Steps Revisited: Identifying *Simulacra* of Holiness

The last sentence of *In His Steps* provides haunting clues about the vision of holiness and community operative in this novel by Charles Sheldon.

> And with a hope that walks hand in hand with faith and love, Henry Maxwell, disciple of Jesus, laid him down to sleep and dreamed of the regeneration of Christendom, and saw in his dream a church without spot or wrinkle or any such thing, following him all the way, walking in His steps. THE END.[6]

The invocation of the theological virtues at the end of this Christian romance belies the fact that the novel provides no social embodiment of these virtues but merely a series of improbable cardboard characters and chimeras of community. Discipleship is present more as a pious yearning than as a set of practices in this story.

Sheldon's story-sermon unfolds in response to the congregation's encounter with a stranger who appears in the Midwestern city of Raymond and whose presence prompts some members of the congregation to rethink their commitment as disciples of Jesus Christ. After spending three days in the city trying to find a job without success, the stranger walks through the entrance of "First Church" on Sunday morning, where he encounters a typical middle-class white Anglo-Saxon Protestant congregation that takes pride in the fact that they have the "best music money can buy" and hears a sermon (based on 1 Pet 2:21) that focuses on the importance of personal sacrifice.[7] After the sermon the stranger stands up in the rear of the church and walks down the center aisle to address the congregation. The stranger tells those assembled that he has been displaced from his job in the newspaper industry by advances in printing technology—the new linotype machines. Then he asks a simple but direct question:

---

[6]Sheldon, *In His Steps*, p. 242.
[7]Ibid., p. 4.

> What do you Christians mean by following the steps of Jesus? I get puzzled when I see so many Christians living in luxury and singing "Jesus, I my cross have taken, all to leave and follow thee,". . . it seems to me there's an awful lot of trouble in the world that somehow wouldn't exist if all the people who sing such songs went and lived them out. . . . But what would Jesus do? Is that what you mean by following in His steps?[8]

Having presented this challenging question, the stranger collapses, and—three days later—dies, but the four-word question that he asked turns out to have an afterlife for many in the congregation.

As Sheldon takes great pain to show his readers, although the congregation was greatly perplexed about what to do, they *listened* to the stranger's plea. Although they were used to encountering such tramps "out on the street," they had "never dreamed" that this kind of event would take place in their church. In the week following the appearance, members were still abuzz about the stranger's appearance in their midst. The following Sunday Pastor Maxwell appears in his pulpit to preach "without notes" (Sheldon's melodramatic way of calling attention to the significance of what was about to take place). His sermon takes up the question of the stranger and makes it the basis for "*a new approach* to discipleship." Maxwell invites "volunteers" from the congregation to take the pledge "to not do anything during the coming year without asking the question 'What would Jesus do?' "[9] And with that invitation, the story unfolds.

The way in which selected members of the congregation of "First Church" proceed to answer that question is cast within a modernist dilemma that separates the practices of discipleship in the present from the practices of first-century Christianity. This puzzlement is articulated early in the novel by Rachel Winslow, the beautiful young woman who displays the yearning to follow Jesus more fully than she had been. On the one hand, the characters of the novel are no longer content with their "previous definition of Christian discipleship."[10] On

---

[8]Ibid., p. 9.
[9]Ibid., p. 15.
[10]Ibid., p. 7.

the other hand, they are uncertain about how to answer the question given the distance between the modern age and the time of Jesus. As one character states: "I am a little in doubt as to the source of our knowledge concerning what Jesus would do. Who is to decide for me just what He would do in my case? It is a different age. There are many perplexing questions in our civilization that are not mentioned in the teachings of Jesus. How am I going to tell what He would do?"[11]

The answer that Pastor Henry Maxwell provides reveals much about the schizophrenia of twentieth-century American Protestantism: "There is no way that I know of . . . except as we study Jesus through the medium of the Holy Spirit. . . . We shall all have to decide what Jesus would do after going to that source of knowledge."[12] In these two sentences the logical contradiction of the novel can be found. On the one hand, the response of discipleship is to be informed by the Holy Spirit; on the other hand, the locus of decision-making is the individual disciple him- or herself. It is significant that neither Rachel Winslow's question nor the pastor's answer recognizes the potential role of the congregation—the gathered believing community—for processing this concern. The combined effect of the modernist sense of separation between past and present and the desire for a clear answer to the question appears to induce a peculiar form of amnesia in which the witness of Scripture, Christian tradition and the practices of the church itself are forgotten, left aside, in favor of a new set of practices that have the appearance of being real but that are largely empty.

"Taking the pledge" then "to not do anything during the coming year without asking the question 'What would Jesus do?'"[13] turns out to be the first—if not the most obvious—*simulacrum* of a Christian practice in the novel. Based as it is on the temperance movement of the late nineteenth century, taking the pledge actually *reduces* Christian commitment to the scale of middle class morality as it was constructed by the leaders of the prohibition movement. Moreover, the

---

[11]Ibid., pp. 17-18.
[12]Ibid., p. 18.
[13]Ibid., p. 15.

pledge that the heroes and heroines of this novel are taking has a termination point; the Rachels and Rollins of the novel are undertaking this "experiment" in discipleship for one year. (It is a convenient manipulation of the plot that the focus shifts to another congregation about the time when the conclusion of the experiment would have come to an end so the reader never confronts the possibility that the pledge would not necessarily be renewed, and therefore the return to the status quo would take place.) The pledge is also voluntary; it is a commitment that people undertake over and above their current form of discipleship, and therefore in some sense turns out to be supererogatory—not a set of obligations that are thought to apply to everyone but only to those who are called to this "higher way." I will have more to say about this concern shortly.

What the pledge appears to *replace* is also significant. Baptism, the outward and visible sign of human transformation, is *not* a practice that takes place in the novel despite the fact that there are persons who are converted. For that matter, other sacraments are also invisible in the novel. In general, there are no visible "means of grace" that enable holiness to take place. In fact, taking the pledge is a voluntary activity the importance of which eclipses the significance of the means of grace, displaces the significance of church membership and thereby *redirects* the focus of Christian discipleship in the direction of "civic righteousness," to use the phrase that several characters in the novel use to describe their individual project of discipleship. I do not think that it is an accident that Sheldon's novel is written at approximately the same time that the last vestiges of "probationary membership" were being left behind in the Methodist Episcopal churches.

In this sense, then, the "after meeting" can be said to provide the *second simulacrum* of the novel. This is the gathering of those persons who have taken the pledge that occurs each week after the regular worship service. This pietistic *ecclesiola in ecclesia* is described at several points in the novel, although it never has more than a shadowy presence. What is clear is that it is a time of "mutual fellowship" where persons report on their experience of attempting to follow "in his steps" as illumined by the question "What would Jesus do?"[14]

What the "after meeting" replaces—or in some sense displaces—is the "giving and receiving of counsel" in the sense practiced in the early Methodist class meetings. Although conversations occur in these meetings after the regular Sunday worship service, and persons are described as experiencing anguish over what they should or should not do, this narrative of personal sacrifice is presented without internal conflict or confusion. In fact, according to the pastor of this fictive congregation, "When it comes to a genuine, honest, enlightened following of Jesus' steps, I cannot believe there will be any confusion either in our own minds or in the judgment of others."[15] The assumption is that the Spirit of God will be the unmediated source of all discernment. Accordingly, the issue of moral and ecclesial authority is effectively displaced by a kind of Christ-Spirit mysticism that is experienced individually. Given this intuitionist emphasis on motive, it is virtually impossible to generate the kind of accountable discipleship that would inquire into the basis of judgments that might have been arrived at in a self-deceived way.

It is in the context of this social setting that the significance of "What would Jesus do?" can best be understood. What work does this question do? Is it more than a *simulacrum*? Potentially yes, particularly insofar as it is asked in the face of the aggregate practices of nominal Christianity. The critical edge of this question certainly works against the self-contented hypocrisy of Christians who by virtue of wealth, social standing and mutual reinforcement are not actually walking in the way of the cross. However, it would be a mistake to believe that the question works without being resourced and/or mediated by other practices. Although it represents a line of inquiry, it does not display practical reasoning as such but rather the absence thereof.

---

[14]Ibid., p. 153. Although the "after meeting" is described most fully midway through the novel, the first time that members of First Church convene following the Sunday morning church service is described on pp. 16-19. See also pp. 36-38 for the second such gathering of would-be disciples.

[15]Ibid., p. 18.

Moreover, the common-sensical clarity of the question belies the fact that the question itself serves the purpose of enabling the pastor and his congregation to escape the discomforts of further questioning, puzzlement and discernment that might unfold if they were participating in the practice of "giving and receiving counsel" where fraternal admonition (Mt 18:15-18) would be practiced.

This point leads me to a further observation about the emptiness of the would-be practice of discipleship found in the novel: those taking the pledge are *not* engaged in the practice of "searching the Scriptures." I think most readers impute this practice to the characters because the language of Scripture is present in the novel in a variety of ways, but in actuality leading characters do not engage in an ongoing conversation with Scripture in such a way that their lives are exposed to the scrutiny of the Word that speaks *over against* themselves. Struggle they do, but they are wrestling with their own guilty consciences as much as anything, and the answers they discover do not include alternatives that would seriously alter their relationship to the growing number of strangers in their midst.

The Myth of Christian America

Written at the end of the nineteenth century at a time when European immigrants—Roman Catholic and Jewish—were coming to America in greater numbers, *In His Steps* is almost entirely populated by Midwestern Anglo-Saxon Protestants. Those characters who do not fit this description are portrayed as dangerous "outsiders" such as socialists and saloonists. The world is filled with problems—"the social problem" and the problem of "the city"—but the unmistakable message of the novel is that these problems can be defeated by a moral *crusade* brought about by individual recommitment in alliance with political reform in "Christian America." This is probably the most seductive and slippery of all *simulacra* of holiness and community to be found in the novel because of the way it conflates civic righteousness and Christian discipleship.

Various characters in the novel yearn for signs of renewal and, like Henry Maxwell, *dream* of a church that will succeed in its "conquest" of the political order in Raymond, in Chicago and in the nation as a whole. Readers of Sheldon's novel are captured by the tantalizing image of what will happen if "Christian America" rises up and takes control of the political process *again.* At points in the novel "Christendom" almost assumes the role of a character as Sheldon calls up his readers to seek "the dawn of a new discipleship" and imagine the "conquering triumph"[16] of Christian America. Such personification is important because "Christian America" is the *implied reader* of much of the novel, such as in the melodramatic scene where the pathetic figure of Loreen (a convert) is borne through the crowd after having been struck and killed by a heavy bottle thrown from a nearby saloon in the Rectangle district of the city.

> Crowd back! Give them room! Let her pass reverently, followed and surrounded by the weeping, awestruck company of Christians. Ye killed her, ye drunken murderers! And yet, and yet, O Christian America! who killed this woman? . . . The saloon killed her. That is, the Christians of America who license the saloon. And the Judgment Day alone shall declare who was the murderer of Loreen.[17]

Loreen, like the stranger who interrupts the service at First Church at the conclusion of Henry Maxwell's sermon, is of course a figure for Christ, the Christ who comes to Christians in America as "one of the least of these" (Mt 25:31-46).

Loreen, like several other characters in the novel, represents Sheldon's attempt to recapture what one character (a bishop) describes as the "lost art" of martyrdom in American Christianity.[18] Virtually all the characters in Sheldon's tale are haunted by the discrepancy between their words and deeds, by the incongruity between the hymns they sing and the lives they lead, and perhaps above all by the intuition that the gospel they proclaim has diverged from the apostolic path precisely because they, like the wealthy church of Laodicea, have become "neither cold nor hot" but "lukewarm" (Rev 3:14-16) in their

---

[16]Ibid., p. 158.
[17]Ibid., p. 116.
[18]Ibid., p. 178.

embodiment of the good news of Jesus Christ.

Some, like the slumlord Clarence Penrose, have visions of Judgment Day, which prompt amendment of life; however, very few of the characters in the novel have such apocalyptic visions. Still, virtually all of them are haunted by their wealth. For example, Pastor Henry Maxwell gives up his summer vacation to Europe. As a result of this conviction of personal self-sacrifice various philanthropic endeavors ranging from the endowment of a Christian newspaper to the founding of a settlement house (to work with immigrants) become the focus of moral action in the novel. In the process these earnest would-be disciples of Christ struggle to obey the mandate of the gospel even as they retain and use their wealth and power for new ends.

Interestingly enough, the mandate for recommitment is *never addressed to all* Christians but is presented as exemplary voluntarism. Even though Jesus' conversation with the rich young ruler is invoked at various points in the novel, something like a *"two-level ethic"* remains despite the fact that there is no basis for this separation in the teachings of Jesus.[19] Individual calling, not corporate solidarity in discipleship, marks this vision of Christian holiness and therefore sharply circumscribes the conception of community. As a result, authentic Christian witness, including but not limited to martyrdom, continues to be displaced—or rather replaced—by the collected set of *simulacra* of heroic *individual* discipleship. And yet, the issue of *personal suffering* for the cause of Christ continues to haunt the imaginations of the characters as if the embodied witness of Christian saints and martyrs ultimately cannot be ignored even if the characters also are unable to make the connection that they find themselves groping toward.

It is this haunted intuition that Sheldon repeatedly addresses, typically in the mouth of Pastor Henry Maxwell, whose final sermon in the novel is clearly addressed to "Christian America."

---

[19]Gerhard Lohfink, *Jesus and Community: The Social Dimension of the Christian Faith* (Minneapolis: Fortress, 1984), pp. 39-40.

> It is the personal element that Christian discipleship needs to emphasize. . . . The Christianity that attempts to suffer by proxy is not the Christianity of Christ. Each individual Christian business man, citizen, needs to follow in His steps along the path of personal sacrifice to Him. There is not a different path to-day from that of Jesus' own times. It is the same path. The call of this dying century and of the new one soon to be is a call for a new discipleship, a new following of Jesus, more like the early, simple, apostolic Christianity, when the disciples left all and literally followed the master. Nothing but a discipleship of this kind can face the destructive selfishness of the age with any hope of overcoming it.[20]

Sheldon's attempt to identify the apostolic past with the future—*without engaging the failures of Christian congregations of the present*—is typical of the fantasy of "the once and future church." As such it provides an example of American nostalgia as it has shaped the practices of popular Protestant Christianity in this culture for much of our nation's history.

Sheldon's story-sermon about what happens when a pastor and some members of his congregation decide to make this pledge to let all of their actions be directed by the standard of *"What would Jesus do?"* (for one year) is a moral tale that has been shaped not only by Protestant Christianity but also by American culture itself. As such it reflects many of the unresolved ecclesiological tensions of the former and the unreflective habits of mind of the latter. And, as I will argue shortly, this also reproduces the Constantinian *simulacra* of Christianity that most commonly takes shape within the fantasy of "the once and future church."

In retrospect it is rather easy to see why this "story sermon," as Sheldon called it, has captured the imagination of millions of Americans over the past hundred years. Although the Social Gospel movement of the late nineteenth century spawned quite a few novels, none of them achieved the success of *In His Steps*. There are several reasons why this is the case.

First, Sheldon's story is *an entertaining romance*. Sheldon origi-

[20]Sheldon, *In His Steps*, p. 237.

nally wrote the story to draw young people to Sunday evening services at Central Congregational Church in Topeka, Kansas, where he was the pastor. Each episode was read from the pulpit, in much the same way that early movies ended "to be continued next week." As such, Sheldon's novel stands as something of a bridge between what Neil Postman has described as "the typographic mind" and "the age of entertainment."[21]

Second, as Ronald White Jr. has observed, Sheldon's novel was influenced by the Social Gospel radical George D. Herron's doctrine of *social redemption through individual sacrifice.* While Sheldon can be seen to temper Herron's radical socialist teachings, he also can be seen to follow Herron insofar as he dreamed of a "Christian state."[22] Like Herron, Sheldon's novel does not rule out coercion; in fact it *presumes* that coercion may have to be used to bring about the reconstruction of society into conformity with the standards of Jesus. The political struggle to take back control of the city of Raymond is but the most vivid example of characters who resolve to "get their hands dirty" by engaging in the political struggle.

Finally, throughout the novel, characters voice the *haunted conscience* of American Protestantism. Two-thirds of the way through the narrative, a bishop is heard to lament that martyrdom is "a lost art," young men and women espouse self-sacrifice in self-conscious imitation of Christian martyrs, and a college president avows that he will overcome his hesitance about getting involved in politics even though he "would sooner walk up to the mouth of a cannon than do this."[23] Wealthy heiresses give away *portions* of their fortune and businessmen take great risks in the course of following "in His steps." At the same time, reassuring indications are provided to the readers that in the disposition of their wealth, no one really expects them to renounce their personal property and become poor like St.

---

[21]See Neil Postman's analysis of the transition from the typographic mind to the Age of Show Business in his book *Amusing Ourselves to Death: Public Discourse in the Age of Show Business* (New York: Penguin, 1985), pp. 59-63.

[22]Ronald White Jr., *The Social Gospel: Religion and Reform in Changing America* (Philadelphia: Temple University Press, 1976), p. 146.

[23]Sheldon, *In His Steps,* p. 88.

Francis.[24] That is to say, this "Christian Romance" (as Sheldon once referred to the novel) is not so much about practices that are socially embodied as it is about self-sacrifice personally defined and determined.

The fact that St. Francis constitutes "the extreme limit" of the imagination for what discipleship involves in the narrative world of this novel is noteworthy. At one level, Sheldon's characters can be seen to struggle against the excesses of utilitarian calculation, but at another level the only kind of prudence displayed in the novel is itself consequentialist. As a result, issues of personal property end up being spiritualized. I suspect that it is precisely because of the remarkable ways in which Giovanni Bernardone of Assisi *embodied* Christian discipleship through renunciation of wealth as well as of knightly dreams of the Crusades, accepting instead the mendicant life in the company of his "Friars Minor" for the sake of the church's mission and ultimately receiving *the stigmata* that he violates the common sense vision of discipleship of the characters in the novel. St. Francis's well-known attempts to *restore and reform* the church and his refusing to dissociate the church of the present from his dreams of what the church might yet be display the antithesis of the fantasy of "the once and future church" displayed in Sheldon's novel.

Pastor Henry Maxwell's vision of the transformation of Christian America (presented in the closing pages of the novel) is worth quoting, precisely because of what is—*and is not*—envisioned.

> He . . . saw certain results with great distinctiveness, partly as realities of the future, partly great longings that they might be realities. . . .
>
> He saw President Marsh of the college using his great learning and his great influence to purify the city, to ennoble its patriotism, to inspire the young men and women who loved as well as admired him to lives of Christian service, always teaching them that education means great responsibility for the weak and ignorant. . . .
>
> And now the vision was troubled. It seemed to him that as he kneeled

---

[24]Ibid., p. 154.

> he began to pray, and the vision was more of a longing for a future than a reality in the future. The church of Jesus in the city and throughout the country! Would it follow Jesus? Was the movement begun in Raymond to spend itself in a few churches like Nazareth Avenue and the one where he had preached to-day, and then die away as a local movement, a stirring on the surface but not to extend deep and far? He felt with agony after the vision again. He thought he saw the church of Jesus in America open its heart to the moving of the spirit and rise to the sacrifice of its ease and self-satisfaction in the name of Jesus. He thought he saw the motto 'What would Jesus do?' inscribed over every church door and written on every church member's heart.
>
> The vision vanished.[25]

But Sheldon does not leave the dream there, for the fantasy of the once and future church demands some kind of resolution, even if it is a resolution that effectively escapes the realities of the present. The next sentence reassures the reader that "It came back clearer than before."[26] This time Maxwell envisions a great convention of "Endeavor Societies all over the world carrying in their great processions at some mighty convention a banner on which was written, 'What would Jesus do?'" Tellingly, Sheldon's vision of Christian renewal depicts a movement that never came to be, but then the romance of "the once and future church" does not aspire to be more than a fantasy.[27]

Notice that only at the end of Sheldon's novel does the figure of Christ become visible. Like St. John, Henry Maxwell has one final vision—not of the New Jerusalem as such—but of Christ himself.

> He saw the figure of the Son of God beckoning to him and to all the other actors in his life history. An Angel Choir . . . was singing. There

---

[25]Ibid., pp. 239-42.

[26]Ibid., p. 242.

[27]Much the same is true of the various "sequels" to *In His Steps* written by Sheldon's imitators. To take but one example, Glenn Clark's novel *What Would Jesus Do?* (St. Paul, Minn.: Macalester Park, 1950), written to celebrate the fiftieth anniversary of Sheldon's book, depicts what happens when Charles Maxwell, the grandson of the character in *In His Steps, reenacts* the pledge of his grandfather's generation in a post-World War II situation of congregational decline.

> was a sound as of many voices and a shout as of a great victory. And the figure of Jesus grew more and more splendid. He stood at the end of a long flight of steps. "Yes! Yes! O my Master, has not the time come for this dawn of the millennium of Christian history? Oh, break upon the Christendom of this age with the light and the truth. Help us to follow Thee All the way."[28]

It is poignant to recall this passage from the last page of *In His Steps* in light of Sheldon's less well-known book *Jesus is Here!* (1914).[29] This later novel attempted to narrate what even the most nostalgic American reader did not find credible. Jesus comes to New York City and confronts the traders of Wall Street! One wonders: Did this sequel not succeed because of the time it was published (the eve of World War I)? Or was it something else, such as the suggestion that if there could be a sequel to the novel, then that in itself would be a signal that the dream had not been realized? These queries are particularly significant given Sheldon's own declared intent in the preface to the sequel: "To bring back to our minds the great fact that Jesus was a very real everyday person. . . . The world needs to know Jesus as a real person, one who walks the earth and is a daily lover of men; that is the main purpose of this story, to bring back to the thought of the modern world the living personality of Jesus."[30]

There is considerable pathos in this pietistic desire for divine reality, but the greatest pathos lies in the fact that the author appears to have lost sight of the fact that the church is the means by which the reality of the gospel is to be conveyed to the principalities and powers (Eph 3:10). To substitute melodramatic visions of the coming Christ for the practices of the church amounts to substituting *simulacra* for the gospel of the incarnation of God in the man Jesus.

The question is whether there might be an alternative to this way of telling the Christian story. I think there is, and to that set of concerns I will turn shortly. But before I do, it might be helpful to call

---

[28]Sheldon, *In His Steps,* p. 242.

[29]Charles M. Sheldon, *"Jesus is Here!" Continuing the Narrative of* In His Steps (New York: Hodder & Stoughton/George H. Doran, 1914).

[30]Ibid., unnumbered page two.

attention to the fact that however modern—and (culturally speaking) American—this saga might be, its roots lie in the fourth century, when the so-called Constantinian shift occurred.

Constantinian *Simulacra* of Holiness and Community

"By this sign, conquer." These are the four words that came to Constantine in his famous dream and the cross is the image that he thought he saw in the sky the next day in that fateful battle at the Milvian Bridge in October 312. But what does the statement mean? It is not immediately clear.

For the first century disciples of Jesus "cross" had a political bearing. The cross, after all, was the form of execution that the Romans used against people who threatened their political hegemony. In Luke 14 Jesus warns the crowd gathered around him that to follow him means they will expose themselves to the accusation that they are causing political unrest.[31] Placing the emphasis on *the sign of the cross,* it is possible to understand "conquer" in ways that subvert the usual associations of power as the sign of "the upside-down kingdom"—to invoke the title of Donald Kraybill's popular study.[32] According to Gerhard Lohfink, renunciation of domination *as a practice* is what the Gospel writers have in mind when they recount Jesus' response to the disciples: "It shall not be so among you."[33] In this sense what it would mean for Christian disciples "to conquer" is highly paradoxical precisely because it has been recast within the context of an apocalyptic perspective brought about by the victory of Christ on the cross. From this perspective, then, notwithstanding what Constantine may have *dreamed* would take place, it is a profound mistake to draw direct linkages between God's will and the military victory over his rival Maxentius at Milvian Bridge.

---

[31]My argument in these paragraphs borrows from John Howard Yoder's essays collected in *For the Nations: Essays Public and Evangelical* (Grand Rapids, Mich.: Eerdmans, 1997), pp. 206-7.

[32]Donald B. Kraybill, *The Upside-Down Kingdom,* rev. ed. (Scottdale, Penn.: Herald Press, 1990).

[33]Lohfink, *Jesus and Community,* pp. 44-50. See also pp. 115-22 for a discussion of this practice in the context of the Pauline correspondence.

But it is also possible to make the cross *instrumental* to particular projects of imperial conquest in such a way that the sign itself is domesticated, and its material content is redefined. There is evidence that Constantine did *redefine* the meaning of the cross. We dare not forget that one of the ways that Constantine's dream was embodied was through the Roman *labarum,* the military standard that heralded the Roman Imperium (in its "Christianized" form) after Constantine. And while the statue that the emperor had erected in the Roman forum depicted him as bearing a cross—"the sign of suffering that brought salvation" according to the inscription he himself provided—other Christians would subsequently carry the cross as their standard as they embarked on the Crusades, the memory of which continues to serve as an unresolved problem for Christianity in the West.

As John Howard Yoder argues, prior to Constantine, "Christians had known as a fact of experience that the Church existed, but had to believe against appearances that Christ ruled over the world. After Constantine one knew as a fact of experience that Christ was ruling over the world, but had to believe against the evidence that there existed a believing Church."[34] In the process holiness and community come to be *redefined,* and the servanthood that the New Testament communities associated with walking in the way of the cross is emptied of virtually all of its material content in favor of the sacralization of particular forms of power whether in the form of the nation-state or in smaller political units where Christians (at least nominally) find themselves in the majority and therefore seek to dictate the direction of change in the name of the Kingdom of God.

What does this Constantinian symbol have to do with the saga of the once and future church? Just this: Constantine did not invent the habit of substituting *simulacra* of Christianity for the actual embodiment of the faith in practice, but by his edicts of toleration and legalization of Christianity he certainly gave permission for a new set of cultural confusions to unfold as Christian practices came to be rein-

---

[34]John Howard Yoder, *The Royal Priesthood: Essays Ecclesiological and Ecumenical* (Grand Rapids, Mich.: Eerdmans, 1994), p. 57.

vented. It is not accidental, then, that the person who postponed his baptism until near the end of his life would be the person who would contribute in a culturally significant way to the transposition of the meaning of the cross and the attendant *redefinitions* of holiness and community. How this transpired is highly complex, of course, but what cannot be ignored is that "after Constantine" practices such as military service came to be regarded as acceptable for Christians, whereas *before* Constantine Christians were not regarded as fit candidates for military service precisely because of the way they embodied Christian holiness in community. The entire practice of the Crusades, within which the practices of military combat, evangelization and baptism came to be conflated, is another example of the ways Constantinian *simulacra* came to be institutionalized, thereby obscuring the practices of holiness and community that characterized the early church. And it is the nostalgic memory of the Crusade that informs Sheldon's version of the saga of "the once and future church."

While Sheldon's novel is certainly more than a Constantinian vision of Christianity, it is also *not less* than Constantinianism. It is a Christian romance constructed within the world of a church defined by the Constantinian alliance. But the question must be asked: Can the genre of the romance provide *the means* for the recovery of the "reality" of Jesus or of Christianity? This is what the Protestant fantasy of "the once and future church" has attempted to do in fiction as well as various narrative histories.

It is noteworthy that within the world of the novel, the newspaper is represented as providing the effective medium for the transformation of the world of Raymond. The church, in this narrative, is not an *Ursakrament*[35] as some contemporary Roman Catholics would think

[35]This technical term used by some European Catholic theologians refers to the way the church makes it possible for persons of faith to be actively incorporated into the mystery of salvation in Jesus Christ through participation in the sacraments. The Flemish Dominican theologian Eduard Schillebeeckx is credited with having introduced this conception of the church as mediator of God's grace in his seminal book, translated into English as *Christ the Sacrament of Encounter with God* (1953). As Geoffrey Wainwright explains, Schillebeeckx presented Christ himself as "the great, original, primary, sacrament *(Ursakrament)* the personal embodiment of the meeting

of it, and not the primary means through which the principalities and powers are transformed as the writer of Ephesians (3:10) represented it. Rather, within the narrative world of the novel the church is little more than the sphere of "spiritual fellowship" *detached* from the means to be used to spread the gospel. And still more oddly, it is the narrative of transformation that comes to be important, as if the story of conversion can be told without ever actually being socially embodied in a congregation.

And this is precisely where the *simulacra* come in, and the confusion begins. To be sure, there are practices in this novel, but they are not recognized as such. Some of these are associated with wealth and economics: First Church has "the best music money can buy" and although Rev. Maxwell gives up his European summer vacation, this is a question of the disposal of discretionary income, not a matter of dire need. Such economic practices circumscribe practices of holiness and community by constructing discipleship in terms that the middle class find acceptable. There are also practices of power associated with elections, majority rule, the administration of city hall, etc. But the power of these practices, taken individually as well as collectively, to shape Christian discipleship is not acknowledged.

In the narrative world of the novel good and evil are depicted within the symbolic complexes of Christendom and "saloonism"; it is a moral vision in which the oppositions are structured by the Temperance movement—the last great Protestant crusade—as well as by the Social Gospel movement that tried to redeem the city from the forces of paganism. But the social configuration that shapes the moral perspective of the novel most determinatively is that of Constantinianism. (No character in the novel can imagine the possibility that Christians might exist, normatively speaking, beyond the precincts of power

between God and humanity. The mystery of God, or God's design for all humanity, had been made flesh in Jesus Christ. The active incorporation of individual human beings into that mystery, or the realization of that design on a widening scale, now takes place in the Church through the sacraments." For more information about this complex discussion in Roman Catholic theological circles, see Wainwright's helpful discussion in his book *Doxology: The Praise of God in Worship, Doctrine and Life* (Oxford University Press, 1980), pp. 70-71.

and control in American culture *as an alternative polis.*)[36] Like Constantinians past and present Sheldon's fictive version of the Christian Coalition is poised to *retake* City Hall and thereby seize the means of political control to root out the moral evils of the world.

It is not conceivable within this Constantinian mindset to think of the church as embodying a "contrast society" in the world.[37] Not surprisingly, while there are small-group meetings for prayer and mutual support of one another's projects of individual response to the question, "What would Jesus do?", catechesis, discipline and accountability in discipleship are *virtually invisible* to readers of this novel. In fact, there is little or nothing that is recognizable as constitutive practices of discipleship. The church that is presented in the narrative world of the novel is little more than an *ecclesiola in ecclesia,* differing from its Pietistic forebears primarily in the relative absence of specifiable practices of discipleship. Here no "watchword" for the day is shared (as the Moravians still practice), no covenant group intrudes to call them to accountable discipleship, and no sacramental piety pervades; there is only *the mystical fellowship* of those individual disciples who hear and respond to the call of the stranger Jesus as an invisible force binding individuals to one another through something like an intuitive web of good motives and right intentions.

As a result, the Christian renewal movement that is reproduced in the novel is inherently unstable precisely because it cannot take any particular social form without collapsing back into the individual motives of would-be disciples. But as a fantasy it does not have to do so because it is little more than a *simulacrum* of discipleship, the appearance of a practice without any real embodiment. The nostalgic

---

[36]Stanley Hauerwas has explored this image of the church in several of his works, most recently in the volume of essays entitled *In Good Company: The Church as Polis* (Notre Dame, Ind.: University of Notre Dame Press, 1995). See also Arne Rasmusson's study *The Church as Polis: From Political Theology to Theological Politics As Exemplified by Jürgen Moltmann and Stanley Hauerwas* (Notre Dame, Ind.: University of Notre Dame Press, 1995).

[37]Lohfink uses the term "contrast society" to describe the social dimension of Christian faith as found in the churches of the New Testament era. See Lohfink, *Jesus and Community,* pp. 50, 55-56, 157-63.

remembrance of "the once and future church" can continue to be reproduced so long as the failure of nominal Christianity exists as its counterpoint.

In part, this response may have to do with American culture itself. We Americans are famously ambivalent about tradition. At the same time we are notoriously uncritical about trying out new ideas. We think that we can *reinvent* almost anything. As the historian and cultural critic Michael Kammen has argued in *The Mystic Chords of Memory: The Transformation of Tradition in American Culture,* "American culture contains a dualistic tension where myths are concerned. We can be iconoclastic but we are much more likely to be permissive and self-indulgent about myth."[38]

Kammen explores the problem of American *amnesia* as a function of "the American inclination to depoliticize the past in order to minimize memories (and causes) of conflict."[39] What this means, in effect, is that very often our cultural memory is transformed and distorted without anyone seeming to notice the difference, and myths often take shape as "traditions" without being acknowledged as such. In sum, over and over again, in the face of ongoing social conflicts about economic power that continue to exist in contemporary American culture, we remember *a past that never was* because it is too painful to confront the challenges of the present. This denial of the conflicted present, Kammen argues, leaves Americans vulnerable to an almost incorrigible nostalgia about our past.

I would argue that Kammen's thesis also applies to the way Christians in American culture remember our past in relation to our images of what the Church should be in the present and future. We continue to try to "reinvent" discipleship and the church without taking into account the ways our practices of Christianity are structured within Constantinian assumptions about the nature of the church that severely limit our conceptions of what Christian discipleship entails. As a result we end up perpetuating the fantasy of "the once and

[38]Michael Kammen, *The Mystic Chords of Memory: The Transformation of Tradition in American Culture* (New York: Alfred A. Knopf, 1991), p. 26.
[39]Ibid., p. 701.

future church" and thereby are distracted from telling the stories of real disciples and congregations where discipleship is actually being embodied.

One way to test the adequacy of Stanley Hauerwas's analysis of the conceptual dislocations of sanctification and the body is to remind ourselves of the inability of most contemporary American Protestants to articulate the social significance of holiness. As I have shown, when we begin to probe the patterns of our wandering in the wilderness of words, we discover the shadowy presence of *the simulacra*—empty conceptual structures—which distract us from engaging in the actual practices of Christian holiness and community.

Practical Reasoning and the Practices of Christian Discipleship

When I assigned Sheldon's novel several years ago as one of the readings in a course I offered on "The Bible and American Culture," I was curious how "generation X" college students would respond to Sheldon's narrative a century after it was written. Perhaps the most interesting response of all was the one offered by "Deborah," one of several earnest, young Protestants who had taking the course that semester. "I understand why it is a flawed novel, and I think I understand why some students would describe it as an 'American romance' and others would think it unrealistic . . . but . . . I just can't escape that question—'what would Jesus do?' Isn't that *the right question,* Dr. Cartwright?" this soon-to-be-graduated college senior asked. In her response, at once earnest and yet devoid of practical reasoning skills, we see the power of Constantinian *simulacra* to direct action when Christian practices are ignored.

As I reflected on Deborah's earnest question, it occurred to me that her response—and that of other students in the class—to Sheldon's novel points up several vexing problems in Christian ethics. Certainly Sheldon's book reveals a great deal about American culture and the power of the culture to shape Christian practices. It might also be said to highlight *some of the limits* of our moral imaginations and perhaps even the failure of institutions of moral education. I have come to believe that each of these concerns is related to the central problem at the heart of Sheldon's novel, the dream of the once and future church.

To renounce amnesia is the *first step* toward the courageous embrace of gospel practices. It may also be the beginning of practical wisdom for most of us who inhabit the congregations of American Protestantism. A *second* step would be to identify what are some of the most constitutive practices for Christian living. Gerhard Lohfink provides a useful starting point in his book *Jesus and Community* (1984) where he identifies a series of practices of discipleship that can be seen in the writings of the New Testament: negative practices like the renunciation of domination, patriarchy and violence along with the positive practices of loving one's enemies, the "praxis of togetherness" and offering hospitality to the stranger, to name but a few. These practices, in turn, can be correlated with virtues—moral excellences—like faith, hope, love, humility, generosity, patience, forbearance, longsuffering, and so on.

Finally, virtues are embodied in the lives of holy people. What some have identified as the medieval "culture of the saints" occurred precisely because moral education was a matter of telling the stories of holy people like St. Francis of Assisi, the rich young man who really did give his riches away to the poor in obedience to the gospel virtues. This is but one of the moral possibilities that are implicitly and explicitly ignored in the pages of *In His Steps*. As the late John Howard Yoder observed, the combined effects of Catholic scholasticism's penchant for universals and the Protestant Reformer's iconoclastic denunciation of abuses has led to an impoverishment of Western morality in so far as the stories of the holy people have come to be largely disconnected from morality.[40]

Sheldon's novel points to the problem of "practical wisdom" and the habits of mind that construct moral problems almost entirely in consequentialist terms, as if the only kind of moral issues that we know how to deal with are dilemmas and the like. On the one hand, the characters of the novel are humbled by their inability to see what

---

[40]Yoder, *For the Nations*, p. 213.

they should be doing; on the other hand, it often seems that they are able to arrive at moral judgments without engaging the Christian tradition at all.

My students and I marveled at the way conversions are narrated in Sheldon's novel. It is as if persons who decide the follow Jesus *already know* what they are supposed to do. There is no post-conversion catechesis, no discipleship training and no apprenticeship to saintly exemplars. Converts like the rich young man Rollin Page are represented as acting independently, single-handedly, even intuitionistically. The novel appears to assume the transparency of the gospel to the individual. Both tend toward a reductionistic mode of thinking that reveals a truncated form of practical wisdom. The naive form that the imitation of Christ takes in Sheldon's novel is an oft-mentioned criticism of his story. But I would argue that the flaw in Sheldon's moral vision is not limited to the simplistic question that provides the catalyst for the plot of the story. More fatal still is the virtual absence of what L. Gregory Jones has described as "formation in moral judgment" among the people of God.[41]

The related problems of impoverished moral imagination and the failures of practical judgment call attention to a less-noticed but perhaps even more telling problem reflected in Sheldon's novel—the question of how to inculcate moral wisdom in young people. Clearly this is a matter that concerned Sheldon because he initially wrote the novel for the young people of his church. But the plausibility of the novel is undercut by the impoverished moral performance of the congregations around us, thereby underlining the fact that this is fiction. This circumstance also renders Deborah's question all the more poignant because from time to time students do want to be *oriented* to the moral life, and they don't know where to turn. Often their questions reveal the embarrassing fact that their instructors are at a loss to give them guidance.

Church-related colleges and universities need to provide the kinds

[41]L. Gregory Jones, *Transformed Judgment: Toward a Trinitarian Account of the Moral Life* (Notre Dame, Ind.: University of Notre Dame Press, 1990).

of courses that would help students distinguish between the *simulacra* of Christianity and the genuine article. Perhaps the novel might even be used as an example for this purpose as I have outlined above. To identify the practices that would need to be in place for authentic renewal to occur would provide guidance about what constitutes embodied holiness for Christians. It would also reassert the importance of moral formation in higher education as well as the necessity of integrating spiritual formation into the curriculum of church-related higher education.

Reengaging the Struggle of Discipleship: Embodying Holiness

Deborah is probably correct. There is something right about the question "What would Jesus do?" but as I have tried to argue in this chapter, the question only becomes productive when oriented within—and contextualized by—actual practices of the Christian faith. When the question is oriented within the context of particular practices, then it is no longer subject to the kind of moral intuitionism that characterizes the actions of the novel, and that too often informs the fuzzy-minded thinking of those wearing W.W.J.D. bracelets, T-shirts and baseball caps.

In the Midwestern city where I live, a United Methodist congregation in a troubled neighborhood was closed at the request of the congregation. The building has been sold to a group of Mennonites who have committed themselves to engage in ministry in that neighborhood despite the fact that virtually none of the members of this congregation originally lived in that area of the city.

Strangers who might walk through the doors of Shalom Mennonite Church typically are met by a group of children who display a diversity of races and backgrounds—instead of the largely homogeneous gathering of suburbanites. This congregation understands that they must embody what it means to be disciples of Jesus. Thus the congregation struggles to discover what living under the sign of the cross means for them. In some sense they experience the burden to embody the sign of the cross each time that they gather.

I think it is probably not an accident that the example to which I am referring is a congregation from the believer's church tradition. According to this approach, the reformation of the church "must be an ongoing process, never finished, because one never gets a grasp of all the errors and all the truth at once."[42] To believe, as John Howard Yoder once put it, that "there are answers that we don't have yet" may be said to be a precondition for participating in practices of searching the Scriptures, giving and receiving counsel, offering hospitality to strangers, etc. This is the basis for a *non-Constantinian* approach to the social embodiment of holiness and community.

Interestingly enough, their presumption (that discipleship must be embodied in order to be proclaimed) is precisely the opposite of that found in the dream of "the once and future church." Instead of engaging in wishful thinking about *what might happen* if the city's newspaper could be taken over, they are engaged in the project of being the church, understood as *the means* by which the principalities and powers will be confronted with the good news of the gospel. So the members of Shalom Mennonite Church tell their children the stories of actual embodiments of Anabaptist martyrs of the past and present an awareness that such narratives—which are often silenced in the consumer culture—can actually render "the sanctified body" as a credible witness. They also train children and adults to engage in simple practices such as the ritual washing of one another's feet (based on the narrative of Jesus' own embodiment of servanthood) in conjunction with their celebration of the Lord's Supper as a way of displaying to one another what it means to embody gospel virtues like peacemaking and humility without self-deception.

But more importantly, as part of their baptismal vows, these Christians have taken vows to "give and receive counsel" from one another, a practice that is foundational for how they understand other practices like Eucharist as well as searching the Scriptures. This is to say that the individual Christian is not the primary arbiter of how she or he will "follow Jesus" but that each must be willing to test the ade-

[42]Yoder, *For the Nations*, p. 157.

quacy of his or her readings of Scripture and discernment of God's will with others in the congregation.

There is nothing romantic about this way of practicing Christianity. This is not the stuff of fantasy precisely because it has refused to succumb to the temptation to forget "the wholeness" of Christian discipleship within which love and justice, purity and practicality, leadership and servanthood are *to be held together* in the remembrance of the real lives of holy people as a means of calling forth new generations to take up the way of the cross. To sustain and inculcate this kind of visionary imaginative discipleship requires a different kind of storytelling from the saga of "the once and future church." Certainly if someone were to write a book about this congregation, it is not likely to sell millions of copies, but I believe that the narrative of what is going on in Shalom Mennonite Church is closer to the real article than the imitation of the real thing that one find in novels like *In His Steps*.

Such practices as "giving and receiving counsel" *should not be alien* to congregations in the Wesleyan/Methodist or Holiness traditions, although it may be more readily recognized when described as engaging in Christian conference. However, as theologian Randy Maddox of Seattle Pacific University has recently observed, given their own strange histories, churches in the Wesleyan/Holiness traditions will likely find that they have to reconstruct the relationship of the *means* and *ends* of sanctification in order to understand the ways that practices such as Christian conferencing are to be enabled by the means of grace. As Maddox has persuasively argued in the case of the Church of the Nazarene, holiness churches have undervalued the role of the means of grace (including the Eucharist) for shaping Christian character in the context of community. This is most noticeable when changes to Wesley's "General Rules" of the Methodist societies are taken into account. For Wesley and the people called Methodists, the set of rules specified under "avoiding evil of every kind" and "doing good in all ways possible" could only be performed as members of the societies were enabled to do so through their participation in the means of grace. By contrast, the Church of the Nazarene "tended to recast the very use of such means into an issue primarily of 'duty.'"[43]

It remains to be seen if Stanley Hauerwas's contentions about "sanctified body" and his proposals for what it means to "characterize perfection" will prove to be persuasive to the heirs of Wesley in American Methodist and holiness congregations.[44] I think it is clear, however, that we can no longer live with the *simulacra* of sanctification as if the formation of a holy people is a matter of individual enactment. The disciplines of Christian living can be formative and transformative when oriented within the context of corporate practices such as searching the Scriptures, foot washing and Eucharist. Only within such contexts as these can we begin to see "what Jesus would do" and discover the light that Scripture can shed on our situation.

Revisiting Sheldon's novel in this era of Promise Keepers rallies, the Jesus Seminar and the ever-popular culture wars reminds us of how prototypically *American* is the church constructed within the pages of *In His Steps*. This is important to observe not only because of the ways *simulacra* of holiness have become confused with actual practices of holiness and community but also because of the manifold ways members of various Christian congregations are confusing the fantasy of the once and future church with what it means to walk in the way of the cross.

The confusions will continue as the twentieth century ends and the third millenium begins. We can be thankful that some of the children who have been given W.W.J.D. bracelets by their Christian elders are wise enough to realize that there is more to Christian discipleship than these tokens convey, but we dare not ignore the fact that they

---

[43]Randy Maddox, "Reconnecting the Means to the End: A Wesleyan Prescription for the Holiness Movement," *Wesleyan Theological Journal* 33 (2): 60. See pages 58-62 for the context of Maddox's argument, summarized here.

[44]Two of the essays in Hauerwas's *Sanctify Them in the Truth: Holiness Exemplified* (Edinburgh: T & T Clark, 1998) explore the problems and possibilities associated with the Wesleyan tradition's emphasis on sanctification. In addition to "The Sanctified Body: Why Perfection Does Not Require a 'Self' (included in this volume), the essay "Characterizing Perfection: Second Thoughts on Character and Sanctification" (pp. 123-42) discusses "what is right and wrong" about Wesley's conception of sanctification in relation to William Law's classic work *A Serious and Devout Call to a Holy Life*.

have yet to be given the kind of (catechetical) orientation to Christian living that can prove to be significant. And if we continue to rely upon surrogates for Christian practices, they will never be oriented in what it means to walk in the way of the cross, or understand the difference between "the upside down" kingdom's imaginative rendering of "By This Sign Conquer" and the Constantinian subversion replicated over and over again in the fantasy of "the once and future church."

We dare not forget that it is only because the God who commands is a God whose grace provides the means of our obedience that we can dare to dream that we might be made holy in this life. It is in this sense that obedience to God's commands understood as "covered promises," as John Wesley would have put it,[45] work *with*—not against—the prudential means of grace. We reconnect sanctification with the body—our own and that of the church as the Body of Christ—when we dare to think of ourselves as persons whose very beings are constituted in the practice of "giving and receiving counsel" as brothers and sisters who walk in the way of the cross.

---

[45]For further clarification of this aspect of John Wesley's theology, see Scott Jones, *John Wesley's Conception and Use of Scripture* (Nashville: Kingswood, 1995), p. 124. As Jones makes clear, the clearest statement of Wesley's hermeneutical principle is found in Wesley's sermon "Upon Our Lord's Sermon on the Mount, V." *Works of John Wesley,* bicentennial ed., ed. Frank Baker (Nashville: Abingdon, 1984), 1:554-55.

7

"And He Felt Compassion"

Holiness Beyond the Bounds of Community

MICHAEL E. LODAHL

One of the distinctive contributions of the postmodern sensibility to contemporary theological reflection is its persistent insistence upon the formative role of specific, identifiably particular *communal traditions* in human being, thinking and doing. To the extent that the modern era, having developed in the light of the Enlightenment, tended to downplay the particularities of place, time and tradition in favor of an idealized and unfettered "reason," the *post*modern recognition of our rootedness in the particularities and peculiarities of peoplehood and tradition—or, we might say with Scripture, in the "boundaries of our habitation"—is an understandable and laudable reaction. We are coming to appreciate that, even in our most speculative and noble flights of thought, we are nonetheless historically and socially situated creatures of flesh and blood, place and time—and *not* disembodied logic-spirits. And because we are situated creatures, we are deeply beholden to those who surround and nurture us in family, society, community, tradition. Such postmodern sensibilities, I believe, are correct, insightful and properly chastening of Enlightenment, modern Western hubris.

Hence, while Kant once wrote that "Enlightenment is man's release from his self-incurred tutelage" and that this tutelage "is man's inability to make use of his understanding without direction from another,"[1] in this time of living in the Enlightenment's shadows we are far more liable to recognize the *utter necessity* of "tutelage" in human thought. In this era widely dubbed the postmodern we are now likely to believe that human beings can only make use of their understanding *from, with* and *for* others; *from* others in the powerful influences of traditions that have preceded and influenced (literally, to "influence" is to "flow into") us; *with* others in the various social contexts that surround and flow into us; and *for* others, particularly the others near to us in our communities of discourse and shared life, in the ever-present temptation to harden our own people's biases and agendas into ideologies.

There is certainly nothing inherently wrong with any of these characteristics of human thinking as analyzed by postmodern sensibilities; in fact, I would argue that they are inevitable given the *radical situatedness* of all human being, thinking and doing. Even the third characteristic mentioned—that of the tendency to think *for* (or in behalf of) those nearest us, those who are "our folk" sharing in our communal traditions—is not only inevitable but a good and admirable thing, if its dangerous and even disastrous extremes can be avoided. Remembering that the word *radical* is derived from the Latin term *radix*, root, when we speak of radical situatedness we are saying, essentially, that human beings are deeply rooted in familial, social, societal, religious, moral, political and ethical traditions of particular times and places—and that, therefore, the profoundly communal and ecological nature of human existence can hardly be overstated. Such observations as these are fairly common in postmodern thinking, which undertakes as one of its primary projects the deconstruction or disso-

---

[1]Immanuel Kant, "What is Enlightenment?" as an accompaniment to *Foundations of the Metaphysics of Morals*, trans. Lewis White Beck (New York: Macmillan, 1989), p. 85.

lution of the "magisterial self": autonomous, sole, independent and bequeathed with its own set of "rights." There is no question that this faulty "self"-concept has dominated Western and particularly North American culture(s) in the last few hundred years; there is equally no question that the sooner this triumphalist notion of the self is undone, the better for us all: indeed, the survival of human beings on this planet may well depend upon it.[2]

Certainly such thinking, though probably not quite so apocalyptically tinged, is an undercurrent of this volume as a whole. The contributors to this book, admittedly in various ways and to varying degrees, are obviously offering a strongly implicit, if not downright explicit, critique of the American holiness tradition's tendency in evangelistic preaching to isolate an individual self and seek its solitary sanctification. This critique, given the state of American evangelical and mainline traditions, is well past due.

I hope I am making clear just how sympathetic I am with such efforts as these to rethink the human: no longer as a solitary subject, but instead as a being-with-others; no longer as autonomous, but instead as drinking deeply from its relations to others; no longer as independent, but instead as interdependent; no longer as objective knower, but instead as situated interpreter and involved doer. Christianly speaking: no longer as solitary, immortal soul to be saved and sanctified, but instead as bodily member of the body of Christ.

There is, of course, a certain kind of ecclesiology that follows: the church is that community of believers that lives by the language of its traditions—a language-*game,* if you will, that shapes its practitioners to see and to experience its projected world in particular ways. Here,

[2]One ought not to offer such judgments, however, without recognizing that this modern Western notion of the autonomous self has been predominantly (if not thoroughly) a *male* self; it truly has been "man." Small wonder, then, that some feminist and other liberationist thinkers have been hesitant to welcome the postmodern dissolution of selfhood, implying as it does a loss of the subject just at the historical moment when so many people are demanding and achieving an appreciable measure of subjectivity. See Luce Irigaray, "Any Theory of the 'Subject' Has Always Been Appropriated by the 'Masculine,'" *Speculum of the Other Woman,* trans. Gillian C. Gill (Ithaca, N.Y.: Cornell University Press, 1985).

instead of the Enlightenment myth of pure and clear access to self-evident truths, the postmodern sensibility suggests that the language of communal tradition "construes" a world into which believers are invited to participate precisely by learning the language and embodying the liturgy and practices of said tradition. George Lindbeck, in his singularly influential text *The Nature of Doctrine: Religion and Theology in a Postliberal Age*, has described this as the "cultural-linguistic" approach, wherein "emphasis is placed on those respects in which religions resemble languages together with their correlative forms of life and thus are similar to cultures. . . . Like a culture or language, it is a communal phenomenon that shapes the subjectivities of individuals rather than being a manifestation of those subjectivities."[3] Those familiar with Lindbeck's categories recognize the notion that he rejects in the previous sentence ("a manifestation of those subjectivities") as representative of what he calls the "experiential-expressive" interpretation of religion and doctrine. Basically, what Lindbeck is rejecting is the notion that there is some common, underlying human *religious experience* that subsequently becomes manifest in different communal traditions; instead, again, a religious tradition (with its intergenerational community, language, rituals and practice) "shapes the subjectivities of individuals rather than being a manifestation of those subjectivities." Thus Lindbeck offers an acute critique of the commonly liberal notion of a generalized and universal religious experience that parallels the postmodern philosophical critique of the Western ideal of a common, universal human reason capable of discerning "self-evident" truths.[4]

All well and good—except that in my opinion, this understanding

---

[3]George Lindbeck, *The Nature of Doctrine: Religion and Theology in a Postliberal Age* (Philadelphia: Westminster Press, 1984), pp. 17-18, 33.

[4]Wilfred Cantwell Smith may provide the best example of a modernist scholar of religions, for he finally reduces all plurality of religious faith and experience down to the category of "faith," an existential but universal attitude finally divorced from any of the historical forms by which it is expressed. See especially his *Towards a World Theology* (Philadelphia: Westminster Press, 1981). Less obvious is John Hick's approach, wherein "the Real" is responded to by human beings and cultures in accordance with the social, religious, political and ethical structures each culture provides. Yet Hick also tends toward a process of "flattening out" religious differences by appeal to the

of the church and its doctrines and practices can all too easily lead to a kind of ghettoization of the church. Is there a way to address our polyphonic postmodern situation—a situation wherein even as we reaffirm and celebrate the specific religious, cultural, political and moral traditions that create and sustain particular communities (and therefore human beings), we also recognize and perhaps even fear the fragmentation, suspicion, xenophobia and, finally, violence that a return to particularity can engender? As the Enlightenment-spawned American experiment of a melting-pot culture appears increasingly to spiral out of control, fragmenting into polarized interest groups and mutually suspicious subcultures, one must wonder whether the church(es) will simply follow suit. Shall we, too, withdraw behind the walls of "our community" and speak almost exclusively to one another in our particular language, the "language of Zion" as "we" interpret it?

It is not that I think anyone, except perhaps the most radical of postmodern advocates, would actually promote such a walled-in mentality as this. On the other hand it is not as though there is no support in the New Testament for such a tendency in ecclesiology: one thinks immediately of the sense of closed community in the Gospel of John, or of Paul's doctrine of the body of Christ and its predominantly "one another" ethics. To be sure, in neither of those cases is the communal circle so tightly drawn as to foster an absolute "us versus them" ecclesial practice. Nonetheless, there is little question that for Paul and John the notion of "community" would correspond rather neatly with the body of Christian believers.

It is my purpose in what follows to advocate another interpretation, equally valid and biblical, of the category of "community"—an interpretation which, I fear, is easily forgotten in the postmodern shuffle. I suspect and hope that it shall provide at least something of

---

category of a unitary and universal salvation mediated through the variety of religious traditions. See John Hick, *An Interpretation of Religion* (New Haven, Conn.: Yale University Press, 1989). For a remarkably incisive critique of Enlightenment modernism's pronounced tendency to postulate a single goal or religious end for all traditions, see S. Mark Heim, *Salvations: Truth and Difference in Religion* (Maryknoll, N.Y.: Orbis, 1995).

a counterbalance to the Lindbeckian accent upon the cultural-linguistic interpretation of religious community—and perhaps even stake a claim for the legitimacy of a sort of "experiential-expressivism" that Lindbeck essentially rejects. It is important, however, in this postmodern era's legitimate demand for a return to the texts and traditions of one's community, that a claim such as this not be rooted in some generalized appeal to a supposedly common human experience but rather in the specific claims of the religious community's holy writ. And so it shall be in this case: the argument that follows is grounded in what I will call a *synoptic relationalism,* i.e., in Jesus' ministering and teaching about God's character and rule as witnessed to in the synoptic gospels. Even more specifically, for the sake of providing focus for this essay, I shall restrict my reflections primarily to Jesus' parable, unique to Luke, about "a Samaritan, who . . . was moved with pity" (10:33).

The Question

Just then a lawyer stood up to test Jesus. "Teacher," he said, "what must I do to inherit eternal life?" (Lk 10:25).

This lawyer, or student of the Torah, may indeed have been testing Jesus, but the question he asked is hardly misplaced or inappropriate to the Jewish tradition. It is in fact highly reminiscent of the Deuteronomic code of blessing: "This entire commandment that I command you today you must diligently observe, so that you may live and increase, and go in and occupy the land that the LORD promised on oath to your ancestors" (Deut 8:1). This Sinaitic theology is predicated on the proposition that God's Torah is a gift of *life,* a blessed way to walk as God's people, graciously offered to the community of Israel. *Do this and you will live* is its fundamental premise. As he nears the end of his life and ministry, Moses proffers this last charge to the people: "I call heaven and earth to witness against you today that I have set before you life and death, blessings and curses. Choose life so that you and your descendants may live" (Deut 30:19).

There is a significant implication in the Deuteronomic theology that apparently underlies the Torah student's question: *What we do*

with our lives really matters. It matters so much that, somehow, our inheriting of eternal life rides on what we *do*. Our Protestant tendency to think of salvation as a matter of *sola fides,* combined with a dualism of inner soul-outer body that even today continues to exercise its influence upon Christian people, makes it difficult for us to appreciate the *bodily* nature of this question. *What shall I do?* implies that it is not sufficient to believe certain things or think certain ways or to undergo certain spiritual experiences; instead, it suggests that the eternal life about which this Torah student asked involves quite seriously how one lives as a body among bodies in a material, creaturely world.

But perhaps, after all, this Torah student is trapped in a Jewish legalism of performance that imagines God to measure us solely by what we do. Where is the righteousness that comes by faith in all of this? Where is divine mercy and the forgiveness of sins? For the moment let us simply acknowledge that Jesus does not reject the question as being inappropriate either to Judaism or to Jesus' own teachings. Perhaps that is why, later in the passage, we find on Jesus' lips words that at least appear to substantiate a Deuteronomic, bodily interpretation of faith: "*Do* this, and you will live" (Lk 10:28) and "Go and *do* likewise" (10:37). And perhaps Jesus will turn out to have been, at least in this passage, not a very good Protestant.

The Commands

[Jesus] said to him, "What is written in the law? What do you read there?" (Lk 10:26).

The Torah student naturally has asked a question appropriate to the Sinai tradition, particularly as rendered by the Deuteronomy code of blessings and cursings; Jesus, in response, replies in the mode appropriate to a Jewish rabbi by posing a counterquestion. It is important to recognize that the Torah student's question implies a community of discourse gathered around a text. It is a question that arises out of the Jewish scholarly community's wrestling with the text of the laws of Moses, and Jesus' reply appeals to the same text. "What is written in the law? How do *you* interpret it?"

The Gospel of Luke, of course, is now a book in the Christian canon of Scripture. It is part of the text around which the Christian community gathers for instruction and direction, for hearing (and, we trust, *doing*) the word of God. Yet when we read from this passage in this Gospel according to Luke, our reading refers us to another text, that of Sinai, and the process of its interpretation. Thus this is a textual witness (Luke's Gospel) to a textual discussion about the laws of Moses; it is a communal discussion, then, among "insiders" in the Jewish tradition but now lodged in the communal reading of the Christian church. Jesus' counterquestion maintains the integrity of the Jewish community even within our Gentile hearing and pursues the question of eternal life in the terms of his Jewish community's text. Jesus need not appeal to authorities or sources beyond that which stands written. He is willing to treat the question within its Jewish communal-traditional context of the Torah. "What is written in the law?"

He answered, "You shall love the Lord your God with all your heart, and with all your soul, and with all your strength, and with all your mind; and your neighbor as yourself" (Lk 10:27).

The student of Torah answers with two commands from the laws of Moses; the discussion, therefore, remains within the bounds of the Jewish community. The first command, from Deuteronomy 6, accentuates the particular and peculiar character of this people singled out by Yahweh: "Keep these words that I am commanding you today in your heart. Recite them to your children and talk about them when you are at home and when you are away, when you lie down and when you rise. Bind them as a sign on your hand, fix them as an emblem on your forehead, and write them on the doorposts of your house and on your gates" (Deut 6:6-8). It is not anybody-in-general who is commanded here to love deity-in-general. Rather it is the uniquely liberated and Torah-gifted people of Israel who are here addressed to love Yahweh, their deliverer from Egypt's bondage and guide to "a land flowing with milk and honey, as the LORD, the God of your ancestors, has promised you" (Deut 6:3).

The second command also issues forth from the Sinai tradition, addressing the issue of how the covenant people ought to live with one another as Yahweh's holy people (Lev 19). For example, they are not to steal, nor deal falsely, nor lie to another (19:11); they are not to revile the deaf nor put a stumbling block before the blind (19:14); they are not to slander nor hate nor take vengeance nor bear a grudge "against any of your people; but you shall love your neighbor as yourself: I am the LORD [Yahweh]!" (19:18). Once more, this command of love for neighbor is addressed *to* the Sinai community and *for* that community's life and well-being as the people of God. The *neigh*bor, in other words, is the one who is *nigh:* the fellow Israelite, fellow Sinai covenant member under the rule of Yahweh.

We know that in the other synoptic Gospels, this citation of the dual command of love for God and neighbor is found on Jesus' lips. In Matthew's Gospel Jesus links the two commands with the observation that the second one "is like unto" the first, suggesting that love for God is inseparable from love of neighbor. He also adds, "On these two commandments hang all the law and the prophets" (Mt 22:39-40). In Mark's version, after Jesus cites the dual command a scribe responds with his approval, "You are right, Teacher; you have truly said that '[God] is one, and besides him there is no other'; and 'to love [God]' . . . and 'to love one's neighbor as oneself,'—this is much more important than all whole burnt offerings and sacrifices" (Mk 12:32-33). Jesus then responds with his own (faint?) praise: "You are not far from the kingdom of God" (12:34).

In all of these synoptic cases God's gift of Torah is to be interpreted through the lens of the two great and similar commands to love God with all one's being and to love one's neighbor as though it were one's very self. This dual love in effect *defines* what it means to be a participant in the community of Sinai. Rabbinic teaching included the challenge to the Jewish people that, no matter how far removed they might become from that revelatory event at Sinai, they were nonetheless to think of themselves as among those who stood at the foot of Sinai and received through Moses the community-creating gift of Torah. This is the way to walk as the people of God in community, the way of life and not death. No wonder Jesus, in responding to the Torah student's citing of the dual command of love, spoke straightforwardly: "You have given the right answer; *do this, and you will live*" (Lk 10:28).

This pair of commands, of course, is important not only to the Jewish community but also to the Christian community as it gathers around its authoritative text. Indeed, it may take on a particular importance for the Wesleyan tradition, given John Wesley's penchant for describing the life of the entirely sanctified believer as a life characterized by this very love of God and neighbor. Wesley too would say, "Do this, and you will live"—for to love God with all of one's being and resources, and to love the neighbor as though it were one's very self, is truly what it means to *live*. This is the life that is, in Wesley's lovely phrase "renewed in love." Such divine renewal compels us increasingly toward the kind of life for which God creates human creatures: a life radically open to God and neighbor, a life of self-giving, other-receiving love. This is the life of "Christian perfection," for in such love the aim *(telos)* of God in creating us is increasingly being fulfilled or perfected *(teleios)*. Given Wesley's understanding of the sanctified life, this Lukan passage becomes all the more critical as a textual site for exploring the nature of such a renewed life.

The Exploration

But wanting to justify himself, he asked Jesus, "And who is my neighbor?" (Lk 10:29).

The question "Who is my neighbor?" is invariably raised from the vantage point of a person within the bounds of community. In fact, we have already suggested that, in terms of the commandment of Leviticus 19:18, it is clear that the neighbor is a fellow Israelite. Scholarly discussions of roughly the time of Jesus tended to extend the circle of "neighbor" no further; indeed, while some were willing to include the Gentile convert or God-fearer under the category of "neighbor," this inclusion was not reached without considerable discussion. The point is, given the theological-textual discussions prevalent during this era, the question "Who is my neighbor?" was not entirely moot. The point of the question, essentially, is to explore the precise limits of the community's boundaries, to discern where to "draw the line" between neighbor and stranger. This is, to be sure, a common practice within communities, for the communal identity often depends upon being able to identify the "other," the "outsider," the "not-us" that dwells somewhat menacingly beyond our communal walls. Just as the Torah student wanted to "justify himself" with this question, so also the "insiders" of a tradition-bound community often seek to justify themselves as somehow superior (stronger, smarter, more righteous, possessing more truth, etc.) over those outside the margins, those "out of bounds."

Yet ironically the holiness code of Leviticus that gave shape and identity to the Sinai community of Yahweh, the commands whereby the Israelites were to be "holy, for I the LORD your God am holy" (Lev 19:2), already included the impetus to reach beyond the limit(ation)s of the Israelite community. It is truly remarkable that Christian commentators and theologians have so rarely noted that the same chapter of Leviticus repeats the command of love—but with a singularly significant difference: "When an alien resides with you in your land, you shall not oppress the alien. The alien who resides with you shall be to you as the citizen among you; *you shall love the alien as yourself*" (Lev 19:33-34)! In other words, the Torah student's question regarding the neighbor was already being enlarged, even exploded, by the Torah itself well before Jesus came along! The Israelites are commanded to love not only the neighbor within the covenantal-communal walls but also the alien outside those walls—and to love all, Israelite and non-Israelite, as they love their very selves!

Jewish scholarship has shown that the Hebrew construction involved in the command to love, whether the object of love be God, neighbor, stranger or even land, means essentially to *do good, to be useful or beneficial,* to the other.[5] Once again the importance of bodily *doing* comes to the fore in our discussion not only of eternal life but now also of love.

[5]See, for instance, Abraham Malamat, "'Love Your Neighbor as Yourself': What It Really Means," *Biblical Archaeology Review* 16, no. 4 (1990): 50-51.

Important as this is, it is equally important to note that the command to love the stranger as one's very self is not simply a command to act in a certain way. It is not simply about developing a certain communal pattern of behavioral practice. Further, the voice of Yahweh does not in this case merely insist that the Israelites treat "outsiders" with love "just because I say so, and I am the LORD!" Instead, the voice of God *makes an appeal to the Israelites' communal-traditional experience:* "You shall love the alien as yourself, for you were aliens in the land of Egypt: I am the LORD your God" (Lev 19:34). The Israelites are to draw upon their own fund of experience of oppression, of being the marginalized strangers, for the internal wherewithal whereby to love the stranger. "Recollect your own experience; remember what it was like to be strangers!" Certainly none but the first generation of exodus people could draw on their own direct experience as such; subsequent generations were instead nurtured by the experience of hearing and bearing the communal stories of their ancestors' oppression, injustice and pain. But the principle remains: recall what it *feels like*—whether in life's bodily experiences or in the bodily experiences of rehearsing the ancestral stories—to be the outcast, the disregarded, the oppressed. Respond to the stranger as though that stranger were you, for that is what you once were. Indeed, just as the people were taught to think of themselves as standing at the foot of Sinai, so the Passover observance was and is a re-living, a re-embodying of the Israelite narratives of Egypt's bitter bondage and Yahweh's gracious deliverance; at this table, the experience of being the stranger is as near as the bitter herbs on one's tongue. Even aside from the Passover meal there are instances aplenty in the Hebrew Bible of this divine appeal to the Israelites' experience of marginalization and oppression, but I will cite only one other, from Deuteronomy:

> For the LORD your God is God of gods and Lord of lords, the great God, mighty and awesome, who is not partial and takes no bribe, who executes justice for the orphan and the widow, and who loves the strangers, providing them food and clothing. You shall also love the stranger, for you were strangers in the land of Egypt. (10:17-19)

This God of gods and Lord of lords is not so high and mighty after all, for this awesome One cares for orphans, widows and strangers. This God provides for their bodily needs of food and clothing, presumably doing so primarily through the covenantal participation of the Sinai community, the Israelite people. Yahweh acts in compassionate love toward the marginalized peoples by appealing to Israelite experience, continually jogging their communal memories of alienation and poverty. Thus again it is not simply obedience to a command that moves the Israelites to "love the stranger," but the divine lure of their own experience.

Such reflections as these obviously anticipate that third character in Jesus' parable, after the priest and Levite, who notices the man lying half dead by the roadside. This character is an "other" to Jesus and his Jewish listeners, an "outsider" in relation to the Jewish community and religious traditions. If we listen carefully to the reading of the Church's holy Scripture, *our* Book, we perhaps shall hear from the lips of *our* Lord words that suddenly explode *our* community and *our* world as construed by *our* traditions.

The Shock

"But a Samaritan while traveling came near him; and when he saw him, he was moved with pity" (Lk 10:33).

Certainly much has been said and written about the three parabolic figures on the road that Jesus trots out across his listeners' minds. It may be worth our while once more, however, to note that the first two figures who "passed by on the other side" are explicitly identified as priest and Levite, as representatives of the temple holiness ideology and practice who, it is easily presumed, avoid contact with the "half-dead" victim because he may in fact be dead already. If this implication is legitimately assumed, then we must acknowledge that Jesus calls into question the ideology of holiness-as-separation that requires that strict boundaries of pure-impure, healthy-sick, inner-outer be maintained. In other words, it is not that these first two characters are somehow evil or heartless, for compassion is not out of the range of their experience. Rather, in fact, the holiness tradition of their community has taken precedence over compassion and dulled their human capacity for feeling-with-another.

Of course, this is what makes Jesus' choice of a third character so remarkable. The Samaritan *is precisely* the outsider, the rejected, the impure, the heretic. We need not rehearse the dynamics of Jew-Samaritan relations of first-century Palestine here, but it is surely not insignificant that there is another passage in Luke's Gospel—one that, like our parable, is unique to Luke and close enough to be our parable's neighbor—that addresses this issue. It is only back in Luke 9:51-55 that we read the story of Jesus and his disciples being refused night lodging in a Samaritan village because "[Jesus'] face was set toward Jerusalem" (9:53). This Samaritan community recognized Jesus and his disciples to be Jews on their way, presumably, to temple worship in the Jewish holy city. Since the Samaritans adamantly insisted that Mount Gerazim was God's appointed site of worship (cf. Jn 4:20), what we find in this passage is a distinct clash between two religious-cultural communities. Is it coincidental that in Luke's telling of the story of Jesus this tale of rejection almost immediately precedes Jesus' own very different tale?

In terms of Luke's narrative sequence, then, the Samaritan rejection of the disciples would be something of a fresh wound. These communities defined themselves oppositionally vis-à-vis one another and as exclusionary of each other. Jesus and his disciples, as Luke tells the story, have had a recent taste of being the rejected "other," of being those disallowed village entry, of being those outcast beyond this community's "walls." And yet now Jesus, safe back "inside" his Jewish community's "walls" of Torah and tradition, reaches beyond those walls to include in his revolutionary parable "a certain Samaritan" who dwells on the outside! It occurs to me that in the very *telling* of this story Jesus was loving the stranger, indeed perhaps loving the enemy, as socio-culturally defined.

And what does Jesus say of this stranger, this outsider, this enemy? He says simply that this Samaritan, this one who walks outside the communal walls, "was moved with pity" (Lk 10:33). He *was moved,* feeling in his *body,* his body permeated by the suffering and vulnerable extremity of the roadside victim. I would suggest that, as Jesus spins this yarn, everything else that this thoroughly unexpected character does for the battered and abandoned man—salving and bandaging his wounds, getting him to an inn, caring for him, paying for his further care—flows from his *being moved with pity or compassion.*

This deserves further probing. Certainly we must beware of overpsychologizing what is, as a matter of fact, only a fictional character in a highly imaginative parable. But the psychologizing element is already present in the description of the Samaritan's response to the man by the side of the road; whereas the priest and Levite "passed by on the other side," at that precise point in the parable's narrative structure the Samaritan "was moved with compassion." Of course, Samaritans shared with Jews the teachings of Moses and thus also were addressed by the Sinai command to love the stranger. But in the parable's own terms as we have received it, it is not Scripture or tradition or community-shaped identity that moves the Samaritan; it is in fact nothing more or less than a human capacity for compassion, for feeling-with "the other," for empathy.

It does not matter whether or not any Samaritan ever did such a thing. It may matter greatly, however, whether or not in this parable Jesus appealed to a human experience—one John Wesley would much later describe as "fellow-feeling"—that first radically undermines our well-defined and boundaried communities and next exposes us to a vastly wider community than we generally imagine. At the very least, this profound and enduring parable in the heart of the Christian canon lends itself to such an explosive interpretation.

The Counter-Question

"Which of these three, do you think, was a neighbor to the man who fell into the hands of the robbers?" (Lk 10:36).

We should recall the Torah student's question that prompted this parable: "And who is my neighbor?" (10:29). By sheer force of habit, it seems, we have come to distill from Jesus' parable our own answer to this question: "Your neighbor is whoever is in need." This too-often trite reply, however, misses the glaring fact that Jesus took the

Torah student's question and turned it entirely inside out! No longer is the question "Who is my neighbor?" Now the question, as Jesus puts it, is "Who proved to *be a neighbor* to the man in need?"

There is more being turned inside out here, though, than the grammatical structure of a question. Recall an earlier point: the question "Who is my neighbor?" is raised from the vantage point of one inside the community, one of privileged place, one who can, in a sense, survey safely from within the walls the almost theoretical question of who's in/who's out. When Jesus turns the *question* inside out, he also shifts the *walls of community* inside out! *Neighbor* is no longer a theoretical figure whose identity is determined by the community's texts and traditions, for suddenly *I* am to *be* the neighbor. The neighbor is the one who walks along the road nigh enough the stranger to *be moved,* to feel compassion, to become the neighbor to the stranger and thus to invite him, likewise, to become nigh.

The Commission

Jesus said to him, "Go and do likewise" (Lk 10:37).

Much hangs, to be sure, on Jesus' final words to this Torah student. We hear again the echoes of the Deuteronomic code, the *Do this and you will live.* Given the nature of Christian tradition as it has developed around the church's Scriptures, such a charge as this may sound a little strange. Is Jesus being merely rhetorical? Does he really expect the Torah student to be able to "go and do likewise"? Does Jesus perhaps harbor grave doubts about anyone's possibility to feel compassion, to show mercy, to be the neighbor, to *do this*? Does such a parting word sound too much like Moses to be at home within the walls of Christian community?

These are not insignificant questions, which may well be the point. In this same Gospel of Luke, we read in Jesus' sermon on the plain such injunctions as these:

> But I say to you that listen, Love your enemies, do good to those who hate you, bless those who curse you, pray for those who abuse you.

> . . . Do to others as you would have them do to you.
>
> If you love those who love you, what credit is that to you? For even sinners love those who love them. . . . But love your enemies, do good, and lend, expecting nothing in return. Your reward will be great, and you will be children of the Most High; for he is kind to the ungrateful and the wicked. Be [compassionate], just as your Father is [compassionate]. (Lk 6:27-28, 31-32, 35-36)

It is *God,* we might say, who proves to be our neighbor, the One who is nigh, who feels compassion for all. For the Most High there is no boundary line, no higher-lower, insider-outsider demarcation. This may become even clearer in Matthew's version of the same passage, where the community Jesus describes is identified only as "children of your Father in heaven [who] makes his sun to rise on the evil and on the good, and sends rain on the righteous and on the unrighteous" (Mt 5:45). Jesus appeals to the common blessings of creation itself as the communal context whereby God's love may be experienced and responded to; this *synoptic relationalism,* accordingly, sees God's neighborly presence and compassionate love in the *community of creation.* This is, in Russell Pregeant's words, "an order of creation perceptible in nature itself and therefore, by implication, available to all persons in all times and places."[6] It is noteworthy, too, that Matthew's version of Jesus' injunction—"Be perfect, therefore, as your heavenly Father is perfect" (5:48)—was a most critical one in Wesley's development of the doctrine of Christian perfection. To be perfect—a rather inadequate translation of the Greek *teleios:* complete, fulfilling one's purpose, "on target"—is to love without boundary. It is, as Luke suggests, to be compassionate and moved with compassion to *do good* (Lk 6:33, 35).

The Ethics of Empathy

Is there in this synoptic relationalism of Jesus, then, the possibility of a religious ethics grounded not in the particularity of Christian com-

[6]Russell Pregeant, *Christology Beyond Dogma: Matthew's Christ in Process Hermeneutic* (Missoula, Mont.: Scholars Press, 1978), p. 78.

munity and tradition but rather in the (potentially) universal human experience of empathy? At least in terms of the Wesleyan theological tradition, I believe that this question can be answered in the affirmative. Drawing upon the work of moral theorist Francis Hutcheson, Wesley in his sermon "On Conscience" postulated that beyond the typical five senses human beings are also possessed of a "public sense, whereby we are naturally pained at the misery of a fellow creature, and pleased at his deliverance from it."[7] Wesley departed from Hutcheson, however, by arguing that this "public sense," or what in other places he calls "fellow-feeling," is not simply an inherent human capacity; rather, says Wesley, it is "a branch of that supernatural gift of God which we usually style, preventing grace."[8] The remarkable point in all this is that, for Wesley, no anthropology is adequate that does not take into account the lively presence and influence of God's prevenient grace in human life and relation—and further, Wesley readily identified this grace with the actual presence of the Spirit of God. To be human is to be graced; to be graced is to be enveloped, enlivened and addressed by the Holy Spirit; and to be so enveloped is to be capable of the "public sense," or what we might call today the experience of empathy! And so, writes Wesley, "[W]e may say to every human creature, 'He,' not nature, 'hath showed thee, O man, what is good.'"[9]

While in one sense the human capacity for empathy is certainly *that*—a variable human capacity—given a Wesleyan anthropology of God's Spirit as a constituting presence in human beings and relations we can say that it is more than merely human. Once again we encounter the notion of God as our closest neighbor, as the One whose presence makes us, keeps us and calls us to be human together—beyond our natural proclivity to create relatively closed communities of discourse and practice. Under the rubric of God's prevenient grace the Wesleyan theological tradition can boldly speak of a larger community, just as Jesus did in the synoptic Gospels; the

---

[7]John Wesley, *Works* 7 (London: Wesleyan Conference Office, 1872), p. 189.
[8]Ibid.
[9]Ibid., p. 188.

whole of creation can be interpreted as God's community, for creation itself is the place where God graciously communes with us and with all creatures. While it would be theologically anachronistic and obvious eisegesis to imply that this doctrine of prevenient grace was what Jesus had in mind in his parable, it is not incorrect to say that, for the Wesleyan interpreter, when Jesus described the Samaritan as moved with compassion he was describing what we mean when we speak of the effects of God's prevenient grace active in human lives and relations.

On the other hand, I suspect in Lindbeck and his followers a hesitation to affirm a belief in the Holy Spirit as God truly present and active in human existence, let alone in the world as God's creation; indeed, in *The Nature of Doctrine* he is reticent to address the doctrine of the Holy Spirit. His comment most pertinent to Wesley's understanding of prevenient grace as the *felt presence of God* is, simply, "The *verbum internum* (traditionally equated by Christians with the action of the Holy Spirit) is also crucially important, but it would be understood . . . [in this model] as a capacity for hearing and accepting the true religion, the true external word, rather than . . . as a common experience diversely articulated in different religions."[10] It seems to me that Lindbeck's model utterly restricts the transformative presence of the Spirit to a particular community's confessional and liturgical language ("the true external word") and practices. Is this the same Spirit who gives life to *all* of God's creatures and "renew[s] the face of the ground" (Ps 104:30)? Is this the Spirit of the God who "is not far from each one of us," since "in him we live and move and have our being" (Acts 17:27-28)?

Similarly, Stanley Hauerwas's approach, as exemplified in his essay in this volume, appears at best to restrict Christ's "real presence" to the church and its sacraments. Though Hauerwas in many ways certainly does revive some crucial Wesleyan themes, I would argue that, in the final analysis, a Wesleyan theology rooted in a confidence in the presence of God's Holy Spirit as prevenient grace, given freely to

---

[10]Lindbeck, *Nature of Doctrine*, p. 34.

God's creation ("I will pour out my Spirit upon all flesh"), can and should always be uneasy with the relatively closed "community and its practices" approach to theological reflection, at least when that community is understood as more or less exhaustive of God's communing presence. One might argue, indeed, that prevenient grace meets and confronts us as Christ's "real presence" in the specificity of bodily encounter with the hungry, the thirsty, the naked, the sick, the imprisoned—in short, with the "stranger" who dwells precariously outside the community's secure walls.

Conclusion

The Wesleyan holiness tradition has tended strongly and, I think, rightly to characterize the nature of God as "holy love."[11] But the term *holy* can very easily erect boundaries; we think again of the priest and Levite in Jesus' story keeping their distance precisely, we listeners presume, for the sake of the temple's ritual holiness requirements. Perhaps in a similar fashion some have employed the adjective *holy* in order apparently to build a theological boundary; often, it seems, the impetus to qualify God's character of self-emptying love with the modifier *holy* is a function of the concern that this divine love should not sound too sentimental or inclusive. A holy love, it might be thought, would tend toward a stern love, a love that may have to keep its distance.[12]

It is helpful, though, to recall that the word-concept *holy* first of all bespeaks utter uniqueness, distinctness, even transcendence. Hence, I would argue that the notion that God is "holy love"—a proper and important designation—ought not to be interpreted so that *holy* is understood as a kind of stern or wrathful qualifer; instead, it should

[11]See, for example, H. Orton Wiley, *Christian Theology* 1 (Kansas City, Mo.: Beacon Hill, 1940), pp. 324, 365-87; H. Ray Dunning, *Grace, Faith and Holiness* (Kansas City, Mo.: Beacon Hill, 1988), pp. 192-204; or Thomas Oden, *The Living God* (San Francisco: Harper & Row, 1987), pp. 98-125.

[12]In much popular preaching and theology, for example, this notion of "holiness" arises in the speculation that God the Father in his holiness "had to turn away from his son" while Jesus bore our sins on the cross—perhaps a handy way to interpret Jesus' cry of dereliction but not very sound theologically.

be interpreted so as to suggest that God's love is utterly unique, totally distinct, in a "class of its own," precisely because it infinitely transcends the limitations of creaturely loves. We human beings so easily ask, "Who is my neighbor?" The God of whom Jesus speaks and whom Jesus embodies, on the other hand, is the Neighbor of all. It is because God's love is a *holy love* that, in the words of the hymn, "God's love knows no limits." This is the boundless perfection of God, and the perfection *(teleios)* for which all humans are created and toward which all are beckoned in and by Christ Jesus. This perfection of love knows no boundaries; it is holiness beyond the bounds of convention and community. To be children of God, says Jesus, is to experience by empathy that all human beings, not to mention all of creation, compose our community of neighbors—and to live accordingly, to "go and do likewise."

8

A Contribution to a Wesleyan Understanding of Holiness & Community

SAMUEL M. POWELL

The themes of holiness and community have a natural attraction for Wesleyans. One would not be venturing too far out on a theological limb with the suggestion that holiness, community and their connection have been among the salient contributions of Wesleyan theology to the Christian tradition. Wesleyans have been at times vigorous proponents of, and at times nearly obsessed with, holiness; John Wesley's innovative and adroit efforts to revitalize Christianity by forming communities of believers in which holiness was strenuously pursued are well-known. However, we will not be unjust to the facts if we observe that Wesleyans may over the years have squandered their patrimony by failing to capitalize on these two foci of their theology, *holiness* and *community*. Accordingly, a renewed effort to grasp their importance and to implement them in the life of Christians is in order.

First we take up the question of holiness. What is needed here is a fresh rethinking of the entire subject, not a dutiful recounting of John Wesley's own understanding of the matter. In spite of being a tireless advocate of holiness and a sensitive witness to the Christian tradition's pronouncements about it, Wesley's own understanding of holiness is not above improvement. In particular, one may take issue with the precise ways in which he and later Wesleyans expounded it, ways that have inclined those within the Wesleyan tradition to connect holiness so closely with ethics that it becomes equivalent to and easily confused with moral improvement. In its most emaciated form, holiness becomes synonymous with conformity to certain behavioral standards whose specificity increases in proportion to their banality. Although those who identify holiness and moral development intuitively and correctly perceive that there is a vital connection between the two, they err in simply identifying them, as though holiness were equivalent to a certain sort of moral character, attainment of which could be empirically verified.

The near identification of holiness with moral development is disclosed in Wesley's choice of certain words to expound the doctrine—words such as *temper* and *affection*. These words have their natural home in a certain type of psychology popular in Wesley's day[1] but which is perhaps less useful today. Recent attempts have been made to resurrect Wesley's approach to holiness using not the language of tempers and affections but instead the language of virtue and character.[2] While one can agree that character is an indispensable component of Christian ethics, it is another question whether a Christian character is to be equated with what the Bible signifies by "holiness." Accordingly, the Wesleyan understanding of holiness, whether in its original eighteenth-century form or in its more recent versions, is missing something of great importance in spite of its many undoubted

---

[1]Randy Maddox, *Responsible Grace: John Wesley's Practical Theology* (Nashville: Kingswood, 1994), pp. 132, 201.

[2]Stanley Hauerwas, "Characterizing Perfection: Second Thoughts on Character and Sanctification," *Wesleyan Theology Today: A Bicentennial Theological Consultation*, ed. Theodore Runyon (Nashville: Kingswood, 1985), pp. 251-54; and *Character and the Christian Life: A Study in Theological Ethics* (Notre Dame, Ind.: University of Notre Dame Press, 1975), p. 194.

strengths. The task with respect to holiness, then, is to distinguish it carefully from moral development while preserving the essential connection between them.

Now we turn to the Wesleyan view of the church. Commentators on John Wesley's theology uniformly draw attention to his genius for fashioning small groups of believers and would-be believers in ways that advanced them in the Christian life. However, it is evident that Wesley had no passionate interest in the theological doctrine of the church.[3] He did not need to waste time developing a theoretical doctrine of the church because his reforming efforts all took place against the backdrop of the Church of England. With the established church responsible for routine priestly functions, Wesley could concentrate on his one passion—the revival of religion in Britain.[4] As a result, Wesleyans did not have to think about the doctrine of the church until Methodism developed into denominations distinct from a national church. Even then the concern was how Wesleyans were to be faithful to John Wesley's vision of Methodism as a renewal movement within the church when they were also seeking to form full-service denominations.[5] It must be admitted that Wesleyans have not yet attained a fully satisfactory resolution of this issue. As a result, there is work to be done in fashioning an

---

[3]In his sermon "Of the Church" Wesley expressly endorsed (in paragraph 16) the Church of England's doctrine of the church. His only comment was to the effect that absolutely pure doctrine is not a necessary condition of being a church (paragraph 19). Having briefly disposed of the theoretical task of defining the church Wesley turned to his customary concern about the proper way of living the Christian life (paragraphs 20-27). *Works of John Wesley,* bicentennial ed., ed. Frank Baker, vol. 3, *Sermons III: 71-114,* ed. Albert C. Outler (Nashville: Abingdon, 1986), pp. 46-57. Wesley's reluctance to delve into the doctrine of the church is also indicated by the fact that in his enumeration of the central core of beliefs that represents the substance of the gospel, there is no mention of the doctrine of the church. Instead this central core is occupied with such "grand, fundamental doctrines" as original sin, justification by faith, the new birth and holiness. John Wesley, *Explanatory Notes upon the Old Testament* (Salem, Ore.: Schmul, 1975), p. ix.

[4]Albert C. Outler, "Do Methodists Have a Doctrine of the Church?" in *The Wesleyan Theological Heritage: Essays of Albert C. Outler,* ed. Thomas C. Oden and Leicester R. Longden (Grand Rapids, Mich.: Zondervan, 1991), pp. 213-14.

[5]Ibid., p. 224; Colin W. Williams, *John Wesley's Theology Today* (New York: Abingdon, 1960), pp. 153-55.

adequate Wesleyan understanding of the church.

The way forward in these matters is indicated by two theologians who stand in a tradition much different from John Wesley's. I refer to Friedrich Schleiermacher and Karl Barth. Barth and Schleiermacher may seem to be unlikely candidates for the role of contributors to Wesleyan theology. Not only do they represent the Reformed theological tradition, a tradition from which Wesleyans have over the years taken great care to distance themselves, but they pointedly differ from each other on significant issues. Nonetheless, Schleiermacher and Barth do indeed have something of importance to say to Wesleyans that will help clarify the nature of holiness and of the church. We will look to Barth for help in understanding holiness and to Schleiermacher for insight into the nature of the Christian community and its relation to holiness.

Karl Barth on Sanctification

Barth presented holiness as a new form of existence.[6] The phrase "new form of existence" may at first suggest that holiness is a matter of ethical comportment, but Barth had something different in mind. Holiness instead consists in our receiving a direction from God. It is God's pointing us toward a new and particular situation. This new and particular situation is not just a different sort of life, as though Christianity consisted simply in moral reformation of one's life; rather, the situation toward which we are pointed is a standing before God. In this standing we become partners with God in a covenant.[7] Holiness, then, is our being addressed by God, our being called by God—and also, through the Holy Spirit, our hearing that call and being affected by it.[8] Note that for Barth holiness is not a possession or a state of being; it instead has a directional character, like a vector. Those who are holy are those who look continually to God as Lord.[9]

---

[6]Karl Barth, *Church Dogmatics*, vol. IV, *The Doctrine of Reconciliation*, trans. G. W. Bromiley, ed. G. W. Bromiley and T. F. Torrance (Edinburgh: T & T Clark, 1956-1969), 2:514.

[7]Ibid., p. 523.

[8]Ibid., p. 526.

[9]Ibid., p. 527.

Elsewhere Barth described holiness as being "opened up" by the Holy Spirit so that we can hear the Word of God.[10] It is not, at least not initially, a doing and certainly does not at first denote an ethical relation to our human neighbors. It is principally the sinner's confrontation with God and being drawn into covenant with God. Not surprisingly, Barth depicted the form of existence that results from this confrontation as our being disturbed and as our being unhappy with the former course of our lives.[11]

Barth reinforced the distinction between holiness and ethical living by emphasizing repeatedly that sanctification is an act of God; it is not at all the sort of thing that we can do for ourselves, either individually or corporately. Humans may, through effort and good will, attain to a developed moral character, but the holiness that Barth described falls on us vertically from God, intersecting the flat plane of our mundane affairs.[12] Barth picturesquely portrayed this act of God's Word as a powerful wind stirring up an eddy in humanity. In this metaphor we humans, drifting along in the stream of our lives, coursing our way along familiar banks, are confronted by a wind from God that comes upon us and disrupts us.[13] Of course, Barth was not unmindful of the fact that this encounter with God happens to real people, living in the midst of social relations that at least in part condition who and what they are. He was also quite sensitive to the effect that this encounter with God should have on those lives and social relations; holiness is no quietistic withdrawal from the world. Nonetheless, he regarded holiness as a mystery and a miracle, something not at all brought about by human activity. Although in the providence of God creatures may prove instrumental in our becoming holy, he insisted that the initial shock of holiness—the calling, the directing—comes from God alone.[14] This difference between what

[10]Barth, *Church Dogmatics*, vol. I, *The Doctrine of the Word of God*, trans. G. T. Thomson, ed. G. W. Bromiley and T. F. Torrance (Edinburgh: T & T Clark, 1936-1956), 1:516.

[11]Barth, *Church Dogmatics* IV/2:524.

[12]Ibid., p. 523.

[13]Ibid., p. 530.

[14]Ibid., pp. 556-57.

God does and the role that creatures may play in holiness is underlined by Barth's opinion that the Word of God, when it comes upon us and calls us, launches a destructive assault on all natural and human institutions because of their implicit if not explicit claim of absoluteness. The orders and institutions that structure our human lives and give them meaning are all objects of this destructive power of God. This power is actualized in the moment of holiness, when those institutions are shown to have merely relative validity and to be subject to the judgment of God.[15]

Those who are holy, who receive this new direction and become disturbed, are called; they constitute the church (the *"ekklesia,"* i.e., "called-out").[16] In hearing this call to obedience the holy are witnesses to the destructive assault of the Word of God on human institutions; they are in fact a sign of this destruction, for this assault has taken place in them and thus becomes a historical event.[17] Accordingly, the church is found whenever humans in the new existence of holiness live in dependence on and are given rebirth by the Word of God, a dependence brought about by the Holy Spirit. The church consists in those who witness the destructive effect of God's word and experience it. Although the church is earthly, human and historical (after all, it is still composed of sinners, even if they are disturbed sinners), its members are witnesses to and signs of God's assault on human institutions because of the vertically falling work of the Holy Spirit.[18]

Two aspects of Barth's concept of holiness deserve special attention. First, it is strictly the work of God and in no way capable of being equated with our moral improvement; second, it consists in being called by God, being lifted, as it were, and set into the presence of God. These two aspects come together in a geometric metaphor, whereby Barth described the call of God to us as perpendicularly intersecting our human life. We humans go about construct-

---

[15]Ibid., pp. 543-44.
[16]Ibid., p. 526.
[17]Ibid., p. 544.
[18]Ibid., p. 616.

ing lives and societies with norms and values and institutions. Yet in all this we traverse the plane of our existence unmindful of God, allowing what we have built up to claim our ultimate allegiance and to establish meaning for us. The Word of God, impressed on us by the Holy Spirit, intersects this mundane plane and calls us into an encounter with the holy God who demands supreme obedience and who judges our self-made lives and institutions. Following this encounter there is much to be said about our conduct in the world and our activity in life and human institutions; this is the subject matter of Christian ethics and Barth distinguished himself by writing voluminously on the subject. But logically (although not always chronologically) prior to our ethical life in the world there is our calling by and encounter with the holy God, a calling and an encounter that constitutes our holiness.

Friedrich Schleiermacher on the Church

It is well-known that Barth's theology differs dramatically from Schleiermacher's in many respects. The difference I wish to focus on may best be understood by drawing attention to the ancient heresies that most irritated Barth and Schleiermacher respectively. Whereas Barth's theological sensitivities were agitated by Arianism and Sabellianism, Schleiermacher feared most the consequences of docetism. Barth wished most to avoid the ancient trinitarian heresies, for in them Jesus Christ falls short of being the full revelation of God; Schleiermacher abhorred the Christological heresy of docetism because it failed to do justice to Christ's full humanity. This comparison reveals to us that Barth was most exercised to preserve the freedom and initiative of God and that Schleiermacher was concerned to show that God's actions are almost always mediated to us in and through natural or human events. Whereas Barth was worried by the potential for emphasizing the concept of mediation to such an extent that faith becomes anthropocentric, Schleiermacher was impressed with the fact that God's grace is transmitted to us in empirically discernable ways. Accordingly, Schleiermacher was much more interested than was Barth in understanding the church as a social institution in which the

Christian life occurs in a guided and somewhat predictable fashion. As a result, Barth's conception of holiness emphasizes God's Word and the work of the Spirit and the absolute priority of God's act to a far greater degree than does Schleiermacher's. Although Schleiermacher granted God's initiative in salvation, the role of mediation is far more central in Schleiermacher's theology than in Barth's.

Schleiermacher's theology contains a powerful understanding of the Christian community that combines an empirical, historical aspect (the community and its structures and institutions) with a transcendent, miraculous aspect (the sinless perfection of Christ that overcomes our sin). The church is a thoroughly human and historical community that nonetheless is the bearer and transmitter of divine grace through its very human structures. The key to grasping his point lies in noting the way in which he opposed the church to the world as two distinct and corporate modes of existence. The Christian's holiness is grounded in the church's "divinely-effected corporate life." Opposed to the church's corporate life is the corporate life of sin and misery that constitutes the world.[19] Having thus distinguished church from world, Schleiermacher went on to observe that in the corporate life that is the world, sin is propagated naturally by means of the network of social relations in which we exist.[20] In the world, considered as a vast corporate life, as one grows up and becomes human by instruction and example and by socialization, one simultaneously becomes sinful—not necessarily morally depraved (since sin is not the same as moral deficiency, just as holiness is not identical to moral development), but ignorant of God and cut off from the spiritual influence of Christ. The world consists in socializing relations that perpetuate themselves over time and inculcate life without God in each individual in every generation. Analogously, the church as a corporate life differs from the world in this respect, that as one grows up in the church one becomes through example, instruction and socialization, holy. Or at least one *may* become holy—

---

[19]Friedrich Schleiermacher, *The Christian Faith*, trans. H. R. Mackintosh and J. S. Stewart (Philadelphia: Fortress, 1976), p. 358.

[20]Ibid., p. 365.

there is nothing automatic or inevitable about holiness. Schleiermacher was thus keenly interested in showing how the divine activity works *through* that which is historical (in this case, the church as a corporate life). God's grace and activity are always mediated to us through natural and historical means, without thereby becoming so totally immersed in history that their divine character is lost. Because of his conviction about mediation, Schleiermacher asserted the impossibility of salvation outside the corporate life of the church, God's grace coming to us only in and through the structures and institutions of the church.[21]

We should note again that there is a great danger here, a danger that Barth never tired of proclaiming, of so conflating God's activity and the human instruments by which God works that they become virtually identical. When this happens, holiness is identified with human progress and becomes something that can be scheduled and programmed by careful human planning.[22] Schleiermacher was aware of this danger.[23] The solution to this problem, he recognized, is a robust and realistic understanding of sin and especially of sin's corporate nature. Because of sin's corporate nature, human community is infected with impulses that direct us away from the knowledge of God. In itself the system of human relations can lead us only into sin. Our sole hope of escaping this network of sin-forming forces is the direct intervention of God's grace. In this way Schleiermacher sought to maintain the crucial difference between church and world and the priority of God's initiative.[24]

Schleiermacher was intent on showing us the concrete ways in which the divine activity is mediated to us. Our holiness is always the work of Christ. However, the mode of Christ's activity today differs

---

[21]Ibid., p. 360.

[22]Stanley Hauerwas's statement that "Christianity is to have one's body shaped, one's habits determined, *in such a manner that the worship of God is unavoidable*" (emphasis added) suggests a version of behavior modification in which the church can establish conditions that will automatically bring about Christian behavior in its members. This seems to make holiness a human work subject to the church's rational planning. See p. 22 in this volume.

[23]Schleiermacher, *Christian Faith*, p. 505.

[24]Ibid., p. 361.

from its historical mode two thousand years ago. Whereas Christ by his teaching and personal action had an immediate effect on the early disciples who saw and heard him firsthand, Christ now works upon us today mediately by means of those who preach his Word.[25] Since his death Christ's personal influence in history has ceased—he no longer acts in the way that individual personalities act; now he is effective through the biblical picture of Christ that is perpetuated in the community through preaching, sacraments and other means, thereby bringing about faith and Christlikeness.[26] How is this possible? Only because in Jesus Christ a power had been introduced into history from God. But it entered history, not universally, but specifically in the corporate life of the church. Having entered history, this power then came to be transmitted to individuals insofar as they are affected by this corporate life. In turn, as they are incorporated into the community, they begin to exert an influence on others so that there is a network of reciprocal relations, each person transmitting this power to some and in turn receiving it from others. As a result the community and its institutions are crucial for holiness for they are the means by which Christ acts today.

In Barth and Schleiermacher we have two sets of concerns. Barth wanted to ensure that we in no way confuse the work of God with any human doing; Schleiermacher wanted to portray the church as a living historical community whose institutions and structures mediate the sanctifying influence of Christ. Barth feared that the liberalizing tendencies of modern theology would make of God no more than a shorthand phrase for human cultural and moral development; Schleiermacher feared a stark supernaturalism that took no account of the historical means of grace through which Christ acts. The question for us is how to do justice to both concerns. The answer lies in the starting point for both Barth and Schleiermacher—Jesus Christ. Jesus Christ is the key to understanding both holiness and the church.

---

[25]Ibid., pp. 490-91.
[26]Ibid., pp. 363-65, 492.

A Barthian Meditation on Jesus Christ and Holiness

The first point to note is that we must derive our understanding of holiness from Jesus Christ; in so doing, we will help ourselves resist the temptation to confuse holiness and human moral development. This is because Jesus Christ is the model of holiness and also because the New Testament writers showed no inclination to represent Jesus as a virtuous man after the fashion of ancient philosophical discussions of virtue. Instead they represented Jesus as being radically open to and obedient to God.

Paul described Jesus' radical openness and obedience in terms of the image of God. For instance, according to Romans 8:29 we have been predestined to conform to the image of the Son of God. Sidestepping the thorny predestinarian language, we observe the purpose of predestination—being conformed to Jesus Christ. Amplification of this notion comes from 2 Corinthians 4:4 and Colossians 3:10. In the former passage Christ is expressly declared to be the image of God; in the latter we are told to put on the new self that is being renewed into knowledge according to the image of the Creator. In Ephesians 4:24 we are enjoined to put on the new self that is made according to God in righteousness and the holiness of truth. This concatenation of texts points us to some connected ideas: renewal, holiness and Jesus Christ, identified as the image of God. According to Paul's theology Christians are to have the same form as Jesus. The new self that is conformed to Jesus is characterized by righteousness and holiness. I am suggesting that having the form of Christ constitutes our holiness—that we are holy as we are conformed to Jesus Christ.

But what does this holiness consist in? In what sense was Jesus himself holy? What was it about Jesus that so impressed the writers of the New Testament? Although the New Testament contains no sustained exposition of Christ's holiness, certain passages do employ striking words that are suggestive. Hebrews 5:7 for example speaks of Christ's reverence (or fear or awe *[eulabeia]*), a characteristic that we are expressly urged to emulate (Heb 12:28). Philippians 2:8 calls our attention to Christ's humility and obedience, while Luke 22:42 emphasizes Jesus' submission to his Father. John 5:19, 30 expands this latter idea by asserting Jesus' continual execution of the Father's will. But aren't these characteristics best understood as virtues? Don't they mark Jesus as the pinnacle of human moral development? No, at least if virtue is understood as the development of human nature through habituation. The New Testament portrait of Jesus is not about what a human can become but rather about a life whose center is God and that is consumed with God's will. The point of all these passages is summed up in the opening chapter of 1 Corinthians in which Paul set forth what has come to be called his theology of the cross—the message that God has chosen weak and foolish and despised things in order to overturn the things considered by the world to be powerful, wise and noble. What distinguished Jesus as wisdom, righteousness and holiness from God (1 Cor 1:30) was precisely his becoming something weak and despised, a person who stood before God in an attitude of utter reverence and obedience, claiming nothing for his own. This theology of the cross, then, together with the cited passages from Hebrews and John's Gospel, gives content to the notion of Jesus as the image according to which we are to be renewed in holiness; it defines the holiness that constitutes the image of Christ. Noteworthy is the fact that Jesus is this image not because of his surpassing virtue, attained through cultivation and habituation, but because of his stark obedience to the will of God. This is not to claim that Jesus was lacking in virtue; however, the New Testament picture of Jesus places no special emphasis on the sort of character development that in the ancient world signaled the attainment of virtue. This is not surprising since virtue in the ancient world signified excellence and accomplishment in being human. Accordingly, virtue is not a notion that is readily suggested by the Bible's teaching about Christ, for the Bible's focus is not on what humans can become through habituation but is instead on being drawn out of ourselves and toward God. In short, while a developed moral character is something of great value and a high good that we should all pursue, it is not the same as being renewed in holiness according to the image of Jesus Christ. This holiness is not so much an emulating of Jesus' virtues as it is a being conformed to Jesus' attitude of reverent submission before God.

At this point a question arises: if holiness is being conformed to the image of Christ, does this all imply that Jesus' own standing before God—his own holiness—consisted in his obedient response to the call of God? One can envision without difficulty the possibility of an adoptionistic Christology lurking in this view, according to which Jesus *became* the Son of God because of his righteous obedience to God. It was to head off just this sort of concern that the ancient church endorsed the doctrine of *anhypostasis,* according to which the human nature of Christ did not exist prior to its being assumed by the second person of the Trinity. As a result, the Christian faith teaches that even Jesus with respect to his human nature possessed nothing that qualified him to be the Son of God. His being the Son of God resulted strictly from an act of God that constituted even his being human. Biblically this point is expressed by the Gospel of John's statements that Jesus did nothing from himself, speaking only what the Father had taught him (Jn 8:28), that Jesus spoke nothing from himself (12:49), and that Jesus had life in himself only because the Father had given it to him (5:26). In brief, even Jesus was totally dependent on God the Father. So to be conformed to the image of Christ or, in a word, to be holy, is to stand like Jesus before God without possession or claim and to receive all the spiritual life that we have from God.

If Jesus Christ reveals the true character of holiness and of the God who calls us into the new life of holiness, then the implications for our understanding of sin are not difficult to draw. Just as holiness is not moral improvement, so sin is not moral failure. Sin instead consists in our failing to be conformed to the image of Christ and in our continuing to live in ignorance of the God who calls into being what does not exist (Rom 4:17) and who chooses the weak and foolish and base in order to overturn the strong and wise and noble (1 Cor 1:27-28). Sin means living heedlessly of this call—either not hearing it or hearing it but refusing to obey.

Paul's letters and John's Gospel confirm this view of sin. According to Paul sin is suppressing the truth in unrighteousness (Rom 1:18), refusing either to honor God or to give thanks in spite of knowing God (v. 21), and failing to have God in knowledge (v. 28). We observe here that sin is never equivalent to being morally deficient or failing to cultivate certain virtues (although we may be sure that the sinners Paul described were in fact morally at fault and deficient in virtue); sin is instead portrayed as a life conducted in willful ignorance of God, an ignorance exacerbated by the fact that the plain truth of God's righteous demands is manifestly present to each of us. Similarly, the letter to the Ephesians understands sin to be a death-like existence in trespasses and a walking according to the world's course of disobedience (Eph 2:1-2). It lies in separation from Christ and living in this world without God (2:12). In this life of sin we are darkened in our understanding, alienated from the life of God, and ignorant and hard of heart (4:18). Holiness, on the contrary, consists in being made alive in Christ (2:5) and in being raised up with Christ and seated in the heavenly places—in the presence of God (2:6). It is, finally, a matter of putting off the old self and putting on the new self (4:22-24). John's Gospel endorses Paul's view of sin by depicting it as loving darkness more than light, as refusing to come to the light, lest one's deeds be exposed (Jn 3:20-21) and as not knowing the light (1:10-11). In this representation sin is again a refusal to approach the light that is God. It is to go our own way and to neglect the call of God to come into the light where there is both judgment and also the presence of God. Contrariwise, holiness must mean to abide in this light and to make it the determiner of our lives. Sin, then, is not so much a failure to live ethically (even though sinners do in fact fail to live as ethically as they ought) as it is a stubborn refusal to acknowledge God and a steadfast fleeing from the presence of God. Holiness, by implication, must consist in our standing before God, in hearing and responding to the call of God into the new life in Christ.

A Meditation on the Church à la Schleiermacher

We have looked to Barth for help in rightly viewing holiness. More, however, is needed, for Christians are not simply individuals confronting God face to face in uncomprehending, traumatic obedience and faith; Christians are also members of a community that conditions their thoughts, experiences and conduct in the world. For aid on this point we return to Schleiermacher and employ his ideas as we explore certain biblical passages.

As we noted previously, the essential point in grasping Schleiermacher's understanding of the church is that the church and the world confront one another as two corporate lives, each propagating a distinctive influence: the sanctifying influence of Christ in the church, the misery-producing influence of sin in the world. Just as Jesus Christ was our point of departure in understanding holiness, so we begin again with Christ in order to rightly apprehend the nature of the Christian community. To this end we focus on the New Testament's portrayal of Christ as the last Adam in 1 Corinthians 15:45-49.

The salient issue in these verses is the contrast between Christ and Adam. Adam was a *living* soul, but Christ is a *life-giving* spirit; Adam was natural, but Christ is spiritual; Adam was earthly, but Christ is from heaven (1 Cor 15:45-47). Already in these verses we get a sense of the contrast, indeed the opposition between the two. This contrast is heightened in verses 48 and 49, which speak of two classes of people: those who bear the image of the earthly and those who bear the image of the heavenly. Although Paul did not here draw out the full implications of this contrast, it is reasonable to assume that he regarded Jesus Christ as the founder of a new humanity, just as Adam was the progenitor of the old humanity. In this view humanity as a whole bears "the image of the earthly." It shares in the characteristics of Adam, especially his sin (Rom 5:19). With a bit of imagination we can see that humanity is here being represented as an ongoing, historical network of relations in which the traits of Adam, its founder, are propagated. As this humanity marches through history, the network of relations ensures that humanity, in its natural existence, continues to exist apart from Christ. In other words, sin is propagated in and through the social relations that constitute the world. While this unfortunate state of humanity *results* in moral weakness and shortcomings, it is not simply equivalent to moral failure and is in fact much more drastic than such failure.[27] It is in fact life without God. The church, on the contrary, is a new humanity, one that bears the image of the heavenly. Just as the first humanity was created in the image of God, so the new humanity is being renewed according to the image of its Creator (Col 3:10). As the first humanity was created by the God who called forth light out of darkness, so the new humanity results from the shining of God's light in human hearts (1 Cor 4:6). But while we may grant that the new humanity is composed of those who are holy, can we also state that this new humanity, as a historical people, has a constituting effect on individual Christians? Yes. This new humanity establishes the context in which individuals come to hear the call of God. Whereas the old humanity consists not only of individuals but also of relations by which individuals are shaped and in which sin is propagated, the new humanity, by analogy, consists both of individuals and of relations by which individuals are formed and conformed to the image of Christ and in which holiness is propagated.

Of course, at this point a problem arises, for while the propagation of sin in the old humanity works inevitably and without fail, the propagation of holiness in the new humanity is anything but inevitable. This fact points us toward the central ethical problem of the Christian faith, namely, that we who are in the church are members not only of the new humanity but of the old humanity as well. Although for us the old humanity lies in the past, its pastness does not nullify its effect on us, for the fullness of the new humanity into which we are called lies in our future. Christian existence, then, is situated between the old and the new, between the past and the future—the past of sin and the future of holiness. Those in the church have heard and responded to the call of God into this new future; however, they continue to live in and among the old humanity and the old humanity's chief effect—the propagation of sin—does not leave them untouched.

What are the consequences of these thoughts for our understand-

---

[27]The relation between sin and moral failure is exhibited in Romans 1, where numerous offenses are seen as the *result* of sin. *Sin* is there portrayed as the refusal to know and honor God.

ing of the church? To mix Paul's metaphors, the church is the body of the last Adam. As a *body,* it has a corporate existence; it is composed of many parts, each exerting a reciprocal effect on the others. Its continued existence in history is sustained by the network of these reciprocal relations. As the body of the *last Adam,* the church is the community of the new humanity; it is the body of those who are holy. As such, it is not only a community composed of holy people, a sanctified community; it is also a sanctifying community. Like the world the church is a community with institutions and structures of socialization; unlike the world, the church is the community of the *new* humanity and its structures and institutions inculcate, not the ignorance of God, but the mindfulness of God—holiness.

However, we must be careful in using the phrase "sanctifying community." It could be taken to mean that the church is a human organization that has been fashioned in such a way that holy people are churned out by it just as automobiles are spewed out by factory assembly lines. The witness of Karl Barth reminds us that holiness is not something that is under our control; it cannot be planned or programmed. The church is not merely an organization whose structure is subject to increasingly rational refinement in the interests of emitting its product more efficiently. Although there is a human—all-too-human—aspect to the church, it also possesses, or better, is possessed by, a divine aspect, a point captured in Paul's depiction of Christ as the head of the body, the church. Consequently, the Christian community is necessary not because it brings about holiness through its organizational structures but because holiness always occurs in a context, and the Christian community constitutes the normal context in which holiness is obtained.

The church, then, is analogous to Jesus Christ; it is an intersection of the divine and the human, the eternal and the historical. As a human community, the church can be studied as a historical and sociological phenomenon. Sciences and disciplines such as sociology and educational theory can be brought to bear to analyze and perhaps improve the way in which the church conveys to its members the life and death of Christ through teaching, preaching and sacraments, and the degrees to which individuals in the church come to have a cognitive knowledge about Jesus Christ. And, of course, there is no end to the useful application of various fields of knowledge in making the church a more effective agent in the formation of character and purveyor of virtue. In short, we can fully acknowledge the utter humanity of the church without detracting from its divine aspect. Just as Christ was in every sense human while simultaneously being fully divine, so the holiness of Christians is entirely understandable both in human terms and also as an act of God.

Concluding Wesleyan Postscript

In the introduction of this essay I mentioned two needs of Wesleyan theology: first, a better understanding of the relation of holiness to ethics so that we avoid both reducing holiness to moral improvement and also separating them unduly; second, a more profound theoretical grasp of the nature of the church and its relation to holiness. I have argued that the thought of Karl Barth and Friedrich Schleiermacher can supply some intellectual support for addressing those needs. It is now time to fashion a contribution to Wesleyan thought about these issues.

We turn first to the relation between holiness and ethics. In the preceding part of this essay I was concerned to distinguish holiness from moral development because of the conviction that Wesleyans have tended to equate the two. Yet while I have been critical of identifying holiness and moral development, I also want to insist that there is an essential connection between holiness and ethics. The task at hand is to specify this connection. Perhaps the best way to describe this connection is to understand ethics as the obverse of sanctification: sanctification is hearing and obeying the call of God to forsake the world (viewed as the system of sinful humanity); ethics is living in the world (regarded as the good creation of God) as a result of hearing that call. Put more simply: the word of God first calls us out of the world and then calls us back into the world. As we are called by God, our world is judged; at the same time, the same world sets the stage for Christian ethics. Before hearing the call of God, we conduct ourselves in the world much like the people of Noah's time who were engaged in eating and drinking and marrying (Mt 24:38). In other words, we carry on with our worldly affairs, which are not necessarily sinful or irreligious. After hearing the call of God it is quite possible, depending on how morally responsible or depraved we have been, that the outward course of our lives will resemble our prior course—or, it could differ dramatically from the prior course of our lives. But now, on this side of hearing the call we are, in Barth's phrase, "disturbed sinners": sinners because we imperfectly hear and respond to the call; disturbed because we now know that our worldly lives are held up for scrutiny to the penetrating light of God's judgment. Now we have seen the glory of God in Christ's face (2 Cor 4:6); we have encountered the God who calls into being what does not exist (Rom 4:17) and who chooses the weak, the foolish and the despised. Before this God our righteous deeds are like filthy rags and we wither away like leaves (Is 64:5). However, just as Elijah, having encountered God in the cave, had then to go forth back to the world and carry out God's command (1 Kings 19), so the Christian returns to the world in order to live as a witness to the call and judgment of God. The ethical life is contained in this return to the world.

We can refine our understanding of holiness and ethics even more clearly. If it is true that ethics is the obverse of holiness, then we must think of ethics and holiness not as two distinct things related as root and trunk but as two forms of the same thing. That is, in order to avoid separating holiness and ethics too much, perhaps it is best to understand them as two forms of the knowledge of God. Holiness is one form of our knowledge of God, a form in which we see God reflected in the face of Jesus Christ. Here we know God as the one who calls into being what does not exist. But ethical living is likewise a form of knowing God. This latter point may not be intuitively obvious because we are not accustomed to thinking of ethics as a form of knowing. However, a brief reflection on the Christian understanding of God will resolve the perplexity. The summit of New Testament teaching about God is found in 1 John 4:8: "Whoever does not love does not know God, for God is love." Although this passage does not give us a formal definition of God, it does for us something even better—it discloses what the knowledge of God is like and, by implication, what it is not like. If God is love, then the knowledge of God cannot be like other sorts of knowledge; knowing God is not like knowing the temperature outside or knowing the genus and species of your pet or knowing your neighbor. This is because God is not a being like any other being; consequently we cannot know God in the ways in which we know other kinds of being. We do not find this matter presented philosophically in 1 John, but it is presented clearly in 4:12: "No one has ever seen God; if we love one another, God lives in us, and his love is perfected in us." This verse indicates that our knowledge of God cannot be separated from love; in fact, it identifies this knowledge with love, first by reminding us that no one has seen God (showing that God cannot be known in the usual ways) and then by connecting the presence of God within us to love. Further, the love that we are discussing is not some feeling of love for things in general but an express and active love for the neighbor (1 Jn 3:15-17).

This meditation shows the connection between the knowledge of God and our ethical conduct in the world. If God is love, then our knowledge of God must also take the form of love. Consequently, ethics is not something added on to our knowledge of God; it is not even a consequence of our knowledge of God. It is, in fact, identical with our knowledge of God, although it is so in a form that differs in principle from holiness. Whereas in holiness we know God by being confronted with God's glory in the face of Jesus Christ, in ethics we know God by loving the neighbor. In short, holiness and ethics are united in love. We may even think of holiness as our obedience to the first great commandment (to love the Lord with all our heart, soul, mind and strength) and ethics as our obedience to the second commandment (to love our neighbors as ourselves).

Next is the understanding of community and its relation to holiness. Wesleyans have historically acknowledged that the church, considered both as community and as institution, is indispensable to salvation. With respect to the church understood as a community,

Wesley assembled both Christians and seekers into classes and bands. As such, the church nurtured people and sought to advance them in the Christian faith. With respect to the church as an institution Wesley affirmed the Church of England's role as dispenser of sacraments and other means of grace. In both capacities the church is necessary for salvation. However, as noted in the introduction, Wesleyan practice of the church has usually outpaced its doctrinal thinking about the church. Today, when Wesleyan practice of being the church has eroded from the standards set forth by Wesley himself, perhaps sustained reflection on the nature of the church will help revitalize the practice.

Two points are appropriate. First, John Wesley's instincts were correct. By drawing on the pietist heritage of gathering Christians into small, nurturing groups and then refining that practice, Wesley introduced something of great importance into the Christian tradition, something that is today being rediscovered and reinvented in a variety of forms. All that is needed to round out this practice theologically is an account of community like Schleiermacher's, in which the biblical image of the church as a corporate body is given precision and sharper intellectual form. This judgment does not constitute an endorsement of Schleiermacher's own theology of the church. It is instead an assertion that Schleiermacher has pointed the way that theologians should follow in the attempt to reflect adequately on the Christian faith and life. What is particularly suggestive about Schleiermacher's view is that by representing the church as a corporate body whose dynamics can be studied rationally he opened the door to a fruitful exchange between pastoral theology and the social sciences. Insofar as the latter study social groups and communities, they may well have great contributions to make toward our understanding of the church as a community. Wesleyans, with a historic interest in the empirical side of religion, can find much here that appeals to their most basic convictions about the character-formative function of the church.

But it is precisely here that I hesitate to fully affirm any premature embracing of the social sciences. While the empirical and social scientific approach to religion has much to commend it, it has one serious liability, at least if it is applied in a certain way. If, as is possible, one defines being a Christian empirically—as having one's habits shaped in Christian ways by the church—then the empirical and social scientific approach will be of supreme importance in our understanding of Christianity. This is because, in this case, what distinguishes Christians from others is their character and the habits from which it arises. This is largely an empirical matter. Further, these habits would be instilled by the church in ways that the social sciences could describe with precision.

My objection to such an understanding of the Christian life and faith is that it would not escape a human-centered circle of thought. It would speak much about the church, but little about God. It would discuss human character and virtue and habits at length but neglect Søren Kierkegaard's sense that the call of God upon us cannot be fully set forth in ethical terms. It would describe carefully the means by which human beings are socialized into becoming Christians, but it would be comparatively uninterested in the subject that consumed Augustine—the mysterious grace of God that comes upon some and not others and that creates in us whatever good we have. It would rightly intend to avoid the modern problem of the self-enclosed, isolated individual self but in place of modern individualistic subjectivity would substitute an equally problematic communal subjectivity.

One would be utterly justified in condemning modern excessive individualism and also in agreeing that humans are essentially social beings who live truly only in and through community. But describing human existence and the Christian life in these social and communitarian terms does not in itself elevate us above the modern world's preoccupation with the human self. A social understanding of Christianity may in fact be merely the latest version of the modern preoccupation with subjectivity. One cannot escape modern human-centered thought by emphasizing repeatedly the significance of human *community*. No amount of stress upon "community" can brings us once again to God-centered life and thought.

What is necessary is to recall Schleiermacher's doctrine of the church, with its twin foci of divinity and humanity, grace and nature, the ideal and the historical. While it is true that the church is a social and historical reality and as such can be studied by the social sciences, it is also true that the church partakes of a divine-human union that makes it analogous to Jesus Christ. We should not regard the church as a vast experiment in social engineering, whose goal is to create a generation of people whose lives exhibit a certain character. Instead we should recall that it is God who works in and through and sometimes in spite of what the church does. Holiness is not an empirical matter. It is not reducible to social scientific explanations and is not caused by community efforts at shaping character. Of course, the church's character-forming efforts must go on; we cannot sit around waiting for people to hear the call of God as though such a thing occurred without the church's preaching and teaching. Yet this character-formation cannot produce this encounter with God; there is a gracious element that defies prediction and control.

I propose, then, that we regard the church as the context in which people come to hear the call of God, not the means by which people are turned into Christians through ecclesiastical versions of behavior modification and social engineering. Schleiermacher's conception of the church can help us to see how the church can simultaneously be a historical institution capable of being studied scientifically and a conduit by which God's grace and the redeeming power of Jesus Christ are given to us. Such an understanding would also help Wesleyans to understand the importance of the church as the living context in which we hear the Word of God and see God's glory in the face of Jesus Christ. This is not to claim that God's Word cannot come to people outside the church—that only people in the *Christian* community hear the call of God. But it is to claim that the church is the community in which the Word of God is heard as Jesus Christ, who is the ultimate revelation of God. Only if we regard the church in this way can we have a theology whose center is God and not the empirical church.

Holiness and community are theological topics of extraordinary importance. This is especially true today as the church adjusts itself to existing in a post-Christian era and strives to define itself over against an occasionally hostile and more often indifferent world. While a beginning has been made in rethinking what it means to be Christian in such a world, much remains to be done. Wesleyans should be at the head of such an effort at rethinking. However, while Wesleyans have often *practiced* rightly, they have not always *thought* deeply. Considerations of the sort mentioned in this essay will contribute to a better *understanding* of the life and faith of Christians in today's world.

Conclusion

Michael E. Lodahl

> We know that the whole creation has been groaning . . . and not only the creation, but we ourselves . . . groan inwardly while we wait for adoption, the redemption of our body. (Rom 8:22, 23)

> I appeal to you therefore, brothers and sisters, by the mercies of God, to present your bodies as a living sacrifice. . . . [For] we, who are many, are one body in Christ, and individually we are members one of another. (Rom 12:1, 5)

> The bread that we break, is it not a sharing in the body of Christ? Because there is one bread, we who are many are one body, for we all partake of the one bread. . . . For all who eat and drink without discerning the body eat and drink judgment against themselves. (1 Cor 10:16-17; 11:29)

It would not be claiming too much to suggest that the essayists of this volume have attempted to offer an extended commentary on the biblical teaching regarding the critical role of *the body*—in all of its various shades of meaning—for the Christian life of holiness. To read Scripture, and especially the writings of Paul, along with our essayists is not to read a disembodied text but instead to peruse a text *of* and *for* real bodies of flesh and bone and blood.

To be a body is to dwell in a bodily, material environment—surrounded by, sustained by, supported by all sorts of other bodies. It is really no surprise that Paul tells us that it is God who gives this great assortment of bodies, that "there are both heavenly bodies and earthly bodies" (1 Cor 15:40), that God gifts creation with a wide array of bodies from humans to animals to birds to fish. Hence, to exist as a human body is always already also to exist in communion, in interrelation, in inextricable *bodily* sociality. One might, then, also characterize the essayists of this book as providing a sustained exploration of the doctrine of Christian holiness (especially as espoused by John Wesley and the Wesleyan tradition) from the vantage point of the postmodern recovery of embodiment and radical sociality. We trust that these explorations have been provocative and fruitful, that they might spark conversation well beyond these pages.

Part of what opens up such conversations is the sense that more could always be said—that in these essays new questions have been raised but perhaps not entirely answered, at least not to every reader's satisfaction. Rather than to summarize the preceding chapters, each of which makes its own case effectively enough, I would like to suggest two possible directions for further reflection and conversation flowing from the general thrust of this volume. One of them begins with the book's dual theme of embodiedness and community and zeroes in on the question of the individual person; the other begins with the same emphases but travels in the opposite direction by expanding the ideas of body and community in more universal and inclusive, and less anthropocentric, directions.

First, the issue of the individual self: it should be obvious that human existence is being interpreted in these pages as fluid, layered, radically contextualized, bodily and thoroughly relational. Does such an interpretation immediately imply the loss of "self"? Will there still be a place for the predominantly Western notion of the "individual"? If there is such a place, how shall this "individual self" be understood? And if not, shall we consider this dissolution, this devolution of the self to be an entirely laudable development in Western thought?

Clearly, the apostle Paul offers in his inspired writings an idea of the human that radically undercuts the self-derived individual of the modern era. Paul writes not simply that Christians are members or parts of the body of Christ but that they are *members of one another* (Rom 12:5). Thus, in their gathered life together in the Spirit they *re-member* not only Christ, they also *re-member, re-compose, re-construct* one another. There is no privatized, individualized, self-composed self with Paul—at least none in the body of Christ. Indeed, even Christ, for Paul, is composed of the bodies of believers—and the parable of the sheep and goats in Matthew's Gospel apparently extends Christ's composition to the hungry, the thirsty, the naked, sick and imprisoned, that is, to those who experience the extremities of bodily vulnerability (Mt 25:31-46). Indeed, in that parable the Son of Man is hardly an isolated, monadic, self-enclosed individual; rather, he is dispersed bodily, needy and vulnerable, among "the least of these" (25:40, 45).

Along these same lines, one might wonder whether Paul's doctrine of the body of Christ owes at least a little to his Damascus road experience: "I am Jesus, whom you are persecuting" (Acts 9:5). Perhaps Paul learned from the crucified, resurrected Christ that truly to be (a "self") is to be fractured, outpoured, dispersed—living, as Paul's disciple Martin Luther would put it many centuries later, "in Christ through faith, in his neighbor through love," and thus "caught up beyond himself into God."[1]

It should be noted, however, that such considerations as these do not preclude Paul or Jesus or Luther—or any of us!—from saying "I" and *meaning something by it.* Paul, for instance, can refer to himself; he can think and speak "I" and can do so at times with considerable confidence (2 Cor 11:22—12:10). In one sense then we find in Paul no loss or lack of individual identity. On the other hand, it is also clear that this self, this "I" that Paul speaks, is not a private, solitary realm. Paul's "self," as it finally turns out to be for all of our selves, is a radically open horizon of social and bodily experience rather than a tightly contoured boundary. It is a self that necessarily includes the other in Christ: "For we do not live to ourselves and we do not die to ourselves" because "we are the Lord's" (Rom 14:7, 8). If, therefore, Paul can think and speak "I," it is an "I" that is "crucified with Christ" such that the simple, solitary "I" no longer lives; rather, the living Christ indwells (and thus opens up) Paul's identity. Here there is, in

---

[1]Martin Luther, *Christian Liberty,* ed. Harold J. Grimm (Philadelphia: Fortress, 1957), p. 34.

the most literal sense, a *con-fusion* (a fusing together) of the bodily/spiritual Paul ("the life I live in the body") and the spiritual/bodily reality of "the Son of God, who loved me and gave himself for me" (Gal 2:20). Christian identity, simply put, is no individual, self-derived "I."

But there is an "I" here, nonetheless! Somehow one suspects that the postmodern fascination with the dissolution of the self cannot be uncritically embraced by people of biblical faith. While certainly the Enlightenment models of the individual self and its "rights" deserve our deepest critical scrutiny, we may also shudder at the alternative possibilities that loom in a world where the dignity of the person is thoroughly dissolved. On the pragmatic level alone, social justice may well demand that a sense of individual identity and dignity be defended. One thinks of the rallying cry of Jesse Jackson (certainly echoing a theme in Martin Luther King Jr.'s preaching) to the African-American people—"I *am* somebody!"—and ought to recognize the dangers implicit in undercutting this hard-fought sense of dignity for people who, for the past several centuries on the North American continent, have been systematically denied a sense of "somebody-ness." There is something possibly quite cynical, even disingenuous, about a philosophical or theological mood that would dissipate the dignity of those who have struggled long and hard—people of color, for instance, or women—to gain the respect due them as creatures in the *imago Dei*.

Here once more we are brought face to face with the stark reality of the body. After all, it is readily arguable that historically the "people of color" and "women" to whom I just referred have been excluded, oppressed and dehumanized *precisely* on the basis of bodily difference. I am reminded of a striking passage from one of the early books of that renegade Baptist theologian Harvey Cox, *The Seduction of the Spirit*, published a good quarter of a century ago. In this passage, Cox is recalling a conversation with the Catholic radical Daniel Berrigan, during which the latter "expressed the theological meaning of flesh with characteristic Berriganian succinctness. 'It all comes down to this,' he said. 'Whose flesh are you touching, and why? Whose flesh are you recoiling from, and why? Whose flesh are you burning, and why?'"[2] To confess that the divine Word became flesh and dwelt among us is assuredly to believe also that *God touches all flesh,* and recoils from none; all creatures have been sanctified, made holy, virtually (or at least potentially) *divinized* by virtue of the Word's participation in creaturely, fleshly and bodily existence. Thus, we move from the plea that we not scuttle the notions of individual identity and agency too readily, happily or simplistically—and move toward a second plea that our notion of "community" not become prematurely closed off from the larger framework of the biblical doctrines of creation and incarnation.

The irony is that from a biblical perspective the body does not signify individual "private property" that "I alone" possess any more than the sense of "self" is privatized or individualized. Think of the words of Hebrews, "Remember the prisoners, as though in prison with them, and those who are ill-treated, since you yourselves also are in the body" (13:3). *Body* here is not that which separates me from the other as if "I" have (or even *am*) "my body" as utterly distinct from "you" and "yours"; indeed, "body" is a shared reality that makes empathy possible, makes *re-membering* of the suffering and the imprisoned possible. In his classic little 1952 study *The Body,* John A. T. Robinson described the Hebraic grasp of embodiedness in this way:

> The flesh-body *[basar]* was not what partitioned a man off from his neighbour; it was rather what bound him in the bundle of life with all men and nature, so that he could never make his unique answer to God as an isolated individual, apart from his relation to his neighbour. The *basar* continued, even in the age of greater religious individualism, to represent the fact that personality is essentially social. . . . The body is that which joins all people, irrespective of individual differences, in life's bundle together.[3]

How different is this Hebraic view from the traditional Western—

[2]Harvey Cox, *The Seduction of the Spirit: The Use and Misuse of People's Religion* (New York: Simon & Schuster, 1973), pp. 218-19.

[3]John A. T. Robinson, *The Body: A Study in Pauline Theology* (Philadelphia: Westminster Press, 1952), pp. 15, 29.

and still reigning—notions of the body! If we could reacquaint ourselves with this idea that the body is a social and ecological system in which we *share,* rather than one of a collection of individual physical *units* set off in privatized isolation from one another, then the Pauline proposition that "we are members one of another" might begin not simply to make sense but to ring true. How much more deeply might our capacity for empathic relations flow if we were schooled in such an experience and understanding of the body! In this way I suspect that the eminently legitimate concerns for social justice toward women and people of color might be addressed more on the grounds of empathic sharing of bodily reality and relations than on the modernist appeal to individual rights. Again it is difficult not to think of Jesus' own resurrection identity as he himself framed it, according to Matthew: "I was hungry and you gave me food, I was thirsty and you gave me something to drink, I was a stranger and you welcomed me, I was naked and you gave me clothing, I was sick and you took care of me, I was in prison and you visited me" (Mt 25:35-36). We who are "members" or bodily parts of Christ's body do indeed *re-member* Christ in eucharistic worship, but to remember Christ is indeed also to remember bodily and empathically "the least of these" in whom Christ is dispersed and with whom Christ is present. It is to remember the otherwise forgotten. "The distinctive characteristic of Christian embodiment," argues Sallie McFague, "is its focus on oppressed, vulnerable, suffering *bodies,* those who are in pain due to the indifference or greed of the more powerful."[4]

Those familiar with McFague's recent work[5] know already that she extends this biblical idea of embodiment-as-compassion to the larger bodily environment in which human beings live; in her words, this sense of the body "ought to include oppressed nonhuman animals and the earth itself."[6] The renewed importance in theology of the

---

[4]Sallie McFague, *The Body of God: An Ecological Theology* (Minneapolis: Augsburg, 1993), p. 164.

[5]See not only *The Body of God* but also McFague's more recent *Super, Natural Christians: How We Should Love Nature* (Minneapolis: Augsburg, 1997), especially chapter 1.

[6]McFague, *The Body of God,* p. 164. For a fascinating theological treatment by John

church as a disciplined and practicing community of Christian disciples, as proper and as justified as it is, ought not to eclipse the broader implications of the doctrines of creation, incarnation and eschatology. For example, perhaps the time has come to reflect anew upon the Johannine claim that "the Word became *flesh* and dwelt among us" (emphasis added), recognizing that the Hebraic word-concept *flesh* refers not simply to the bodily existence of human beings but of all creatures (e.g., Gen 9:9-16). The Christian confession that the divine Word has become flesh means, then, that God in Christ shares not simply in human existence but in *creaturely existence* in general. One cannot help but think of Irenaeus's eucharistic reflections in *Against Heresies:*

> For if this [flesh] is not saved, then neither did the Lord redeem us by his blood, nor is the cup of the Eucharist the communion of his blood, and the bread which we break the communion of his body. For blood is only to be found in veins and flesh, and the rest of human nature, which the Word of God was indeed made [partaker of, and so] he redeemed us by his blood. . . . For since we are his members, and are nourished by creation—and he himself gives us this creation, making the sun to rise, and sending the rain as he wills—he declares that the cup, [taken] from the creation, is his own blood, by which he strengthens our blood, and he has firmly assured us that the bread, [taken] from the creation, is his own body, by which our bodies grow.[7]

Blood and wine, bread and body, veins and flesh, sunshine and rain—Irenaeus employs these evocative images to excite in his readers the rich textures of a Christian soteriology that includes the bodily materials and processes of creation. He reminds us that in the Eucharist we share in bread made from grain that, nurtured by sunshine and rain, grows from the earth from which we ourselves come and

---

Wesley of the sufferings of nonhuman creatures and the Christianly human responsibility toward those creatures, see his sermon based on Romans 8:19-22, "The General Deliverance," in *The Works of John Wesley,* 3rd ed., 14 vols. (reprint of 1872 ed., Kansas City: Beacon, 1978-1979), 6:241-52.

[7]Irenaeus *Against Heresies,* in *Early Christian Fathers,* trans. and ed. Cyril C. Richardson, Library of Christian Classics 1 (Philadelphia: Westminster Press, 1953), p. 388.

upon which we stand; that same bread bodies forth to us the body of Jesus Christ. By the same token, the cup's rich red liquid flows from creation's bounty to nourish our blood. All of these rich blessings flow from God the Creator through the incarnate Christ, who sits together with us at the table—a table that celebrates both creation and redemption in the "tasty grace" of the sacrament! It is not at all difficult to argue, with the aid of Irenaeus, that the simple act of participation in the Eucharist has profound ecological implications. The words of Methodist theologian and ethicist L. Harold DeWolf are pertinent:

> When the divine Word was made flesh, . . . God became incarnate not only in human flesh, but in the world's flesh. Jesus Christ . . . was of the same flesh with men and women, yes, but also with deer and fish, with trees and plankton, with seas and mountains. . . . His very breath participated in the oxygen carbon-dioxide cycle by which all plants and animals enable each other to live. By being a part of the web of life Jesus Christ has sanctified not only humanity but the whole fabric of life in which human life exists. . . . Jesus took bread and wine in his hands and said, "This is my body; this is my blood." We must now learn that these words actually implied also *all soil and life, all streams and oceans, "This is my body; this is my blood."* For so they are—his and ours.[8]

Certainly other questions and issues shall flow from this book as themes for further conversation, but I propose that one of the most important goals challenging postmodern Christian theology is the wedding of the two concerns I have attempted to explore in this concluding chapter: on the one hand, a sense of the self, with its identity, dignity and accountability; and on the other hand, a sense that this self is so much larger than modernism's individual ego, that it is in fact intimately, internally and *bodily* connected with many overlapping communities, human and nonhuman. It is no easy task to maintain both of these senses and do them their proper justice—and yet I

---

[8]L. Harold DeWolf, *Responsible Freedom: Guidelines to Christian Action* (New York: Harper & Row, 1971), p. 250.

am convinced that nothing less is demanded of us who live by the conviction that "in Christ God was reconciling the world to himself, not counting their trespasses against them, and entrusting the message of reconciliation to us" (2 Cor 5:19).

List of Contributors

The essays in this volume were originally presented at the Wesleyan Center for 21st Century Studies (Point Loma Nazarene University) February 1998 Writer's Symposium. The symposium engaged essayists with the Wesleyan theme of holiness in a pluralistic culture, as articulated by Stanley Hauerwas at "What Happens to 'Person' in a Post-modern Era?" a January 1997 conference sponsored by the Wesleyan Center.

Michael G. Cartwright is an assistant professor in the department of philosophy of religion at the University of Indianapolis. His publications include contributions to *Disciplines and the Discipline; The Ethics of War and Peace* and *The Gospel in Black and White: The Role of Theology in Racial Reconciliation*. He is also the editor of *The Royal Priesthood: Essays Ecclesiological and Ecumenical,* a collection of essays by John Howard Yoder, and *The Hauerwas Reader.*

Rodney Clapp is the editorial director of Brazos Press. He is the author of *Families at the Crossroads* and *A Peculiar People,* coauthor (with Robert Webber) of *People of the Truth* and editor of *The Consuming Passion*. He has written over one hundred articles for magazines and journals.

Joyce Quiring Erickson is the director of general education and faculty advising and a professor of English at Seattle Pacific University. Recent publications include "'Perfect Love': Achieving Sanctification as a Pattern of Desire in the Life Writings of Early Methodist Women," *Prose Studies* (1997); "Parker Palmer's *To Know As We Are Known* Twelve Years Later," *Christian Scholar's Review* (1996); and an essay ("Homecoming") in *Rattling Those Dry Bones: Women Changing the Church*.

Stanley M. Hauerwas is Gilbert T. Rowe Professor of Theological Ethics at Duke Divinity School. Among his many books are *The Peaceable Kingdom; A Community of Character* and (with Will Willimon) *Resident Aliens: Life in the Christian Colony*. His more recent books include *Christians Among the Virtues; Wilderness Wanderings: Probing Twentieth Century Theology;* and *Sanctifying Them for the Truth: Holiness Exemplified*. He has been selected to deliver the prestigious Gifford lectures in 2000-2001.

Craig Keen is professor of theology at Olivet Nazarene University. His publications include two articles in the *Wesleyan Theological Journal*: "*Homo Pre-*

carius: Prayer in the Image and Likeness of God" (1998) and "(The) Church and (the) Culture: A Little Reflection on the *Assumptio Carnis*" (1989).

Michael E. Lodahl is professor of theology at Point Loma Nazarene University. His books include *Shekinah/Spirit: Divine Presence in Jewish and Christian Religion* and *The Story of God: Wesleyan Theology and Biblical Narrative.* He has contributed several articles to the *Wesleyan Theological Journal.*

Samuel M. Powell is professor of philosophy and religion at Point Loma Nazarene University. He is the author of *The Trinity in German Theology.*

Theodore Runyon is professor of systematic theology (emeritus) at Emory University. He is the author of *The New Creation: John Wesley's Theology Today* and editor of *What the Spirit Is Saying to the Churches; Hope for the Church: Moltmann in Dialogue with Practical Theology; Sanctification and Liberation: Liberation Theologies in the Light of the Wesleyan Tradition; Wesleyan Theology Today* and *Theology, Politics and Peace.*